DUCK PRODUCTION AND MANAGEMENT

This book covers nearly all the aspects of duck rearing system with various duck genetic resources of India, their nutritional requirements, care management, diseases and prevention, slaughter, grading & packaging of meat, nutritional importance of meat & egg and integrated farming. To make this book more competent for farmers, enlisted certain drugs for treatment of different diseases of duck and vaccination schedule has also been included in the text. References have been included in this book to refer the interested reader for details understanding of duck rearing system. The language is kept simple and lucid for ease of understanding. This book will find right place in the hands of researchers, veterinary graduates, students of the animal sciences and farmers who are engaged in duck farming.

Jowel Debnath is presently working as Assistant Professor in the Department of Instructional Livestock Farm Complex (Animal Genetics & Breeding), College of Veterinary Sciences and A.H., R.K.Nagar, West, Tripura-799008 since 2012. He obtained B.V.Sc & A.H degree from College of Veterinary Sciences and A.H., Selesih, Aizawl, Mizoram (India) in 2009. He did his M.V.Sc and Ph.D in Animal Genetics & Breeding from Division of Animal Genetics, Indian Veterinary Research Institute, Izatnagar, Bareilly, U.P-243122 in the year of 2011 and 2016, respectively. He cleared ICAR-ASRB NET in the discipline of Animal Production and Animal Genetics & Breeding in the year of 2012 and 2016, respectively and also cleared the ICAR's All India Examination, ICAR-SRF (PGS)-2011-12. He was awarded for IPSA Prof. P.K. Pani research award-2015 from Indian Poultry Sciences Association, ICAR-CARI, Izatnagar, Bareilly, U.P-243122. The author is actively engaged in research, extension and teaching in undergraduate level. So far, he has published more than 25 research paper, review and popular articles in reputed national and international journals.

DUCK PRODUCTION AND MANAGEMENT

Edited by
Dr. Jowel Debnath
M.V.Sc & Ph.D (Animal Genetics & Breeding)
Assistant Professor,
Department of Livestock Farm Complex
College of Veterinary Sciences & A.H., R.K.Nagar, West,
TRIPURA-799008 (INDIA)

CRC Press
Taylor & Francis Group
Boca Raton London New York

CRC Press is an imprint of the
Taylor & Francis Group, an **informa** business

NPH

NARENDRA PUBLISHING HOUSE
DELHI (INDIA)

First published 2023
by CRC Press
4 Park Square, Milton Park, Abingdon, Oxon, OX14 4RN

and by CRC Press
6000 Broken Sound Parkway NW, Suite 300, Boca Raton, FL 33487-2742

CRC Press is an imprint of Informa UK Limited

British Library Cataloguing-in-Publication Data
A catalogue record for this book is available from the British Library

ISBN: 9781032388878 (hbk)
ISBN: 9781032388892 (pbk)
ISBN: 9781003347293 (ebk)

DOI: 10.4324/9781003347293

Typeset in Arial, Calibiri, Cambria, Helvetica, Symbol and Times New Roman
by Amrit Graphics, Delhi 110032

CONTENTS

PREFACE

Management of animals and birds is both an art and a science. Any livestock enterprise needs sound application of basic scientific principles to achieve maximum productivity and profitability. Management of livestock/ poultry requires the comprehensive knowledge on available genetic resources, feeding, housing and disease control in such a way appropriate for particular situation and region. Proper management at right time in proper way will help to become a successful farmer or manager.

The natural resources of our country have tremendous potential for development of duck farming. Due to deficient in proper guidance or scientific farming protocol on duck rearing farmers can't ensure the proper earning in the country. Therefore, it is more essential to make available scientific information on duck rearing for the farmers. Hence, a small effort has been made to highlight on basic principles of rearing of duck. The objective of this book is to put in the picture of duck rearing in modern scientific way to ensure greater productivity and earning of the farmers. In this book we have gathered essential information in different aspects on the subject of duck rearing, which could be beneficial for the farmers' community of India.

The information of this book could be of use for the students, researcher and any person willing to know about duck genetic resources of India, duck nutrition, duck management, duck diseases, slaughter of duck and integrated duck cum fish farming.

As the editor of the book, I have the pleasure of expressing my sincere gratitude and appreciation to all the contributors without whose help and efforts, it would not have been possible to write this book.

I hope and believe this book would serve as valuable practical implementation in duck farming in our country.

Jowel Debnath

LIST OF CONTRIBUTORS

Dr. Jowel Debnath
M.V.Sc & Ph.D (Animal Genetics & Breeding)
Assistant Professor,
Department of Livestock Farm Complex
College of Veterinary Sciences & A.H., R.K.Nagar, West,
Tripura-799008 (India)

Dr. Bikas Chandra Debnath
M.V.Sc & Ph.D (Animal Nutrition)
Assistant Professor, Department of Animal Nutrition
College of Veterinary Sciences & A.H., R.K.Nagar, West,
Tripura-799008 (India)

Dr. Tapan Kumar Das
M.V.Sc & Ph.D (Animal Nutrition)
Assistant Professor,
Department of Livestock Farm Complex
College of Veterinary Sciences & A.H., R.K.Nagar, West,
Tripura-799008 (India)

Dr. Biplab Debroy
M.V.Sc & Ph.D (Veterinary Pathology)
Assistant Professor,
Department of Veterinary Pathology
College of Veterinary Sciences & A.H., R.K.Nagar, West,
Tripura-799008 (India)

Dr. Bijoy Kumar Sarkar
M.V.Sc & Ph.D (Livestock Products Technology)
Assistant Professor,
Department of Livestock Products Technology
College of Veterinary Sciences & A.H., R.K.Nagar, West,
Tripura-799008 (India)

Dr. Debajyoti Sarkar
M.V.Sc (Animal Reproduction, Gynaecology & Obstetrics)
Ph.D Scholar, Division of Animal Reproduction,
Gynaecology & Obstetrics
ICAR-National Dairy Research Institute,
Karnal, Haryana (India)

CHAPTER 1

DUCK GENETIC RESOURCES OF INDIA

DR. JOWEL DEBNATH

1.1 INTRODUCTION

In India, poultry reared mostly for meat and egg purpose. Among the various poultry species, duck is having a special place. Duck is a water loving bird reared by the farmer since century under traditional system with local low productive non-descript breed. Ducks have ability to lay more egg, larger egg, require lesser attention and thrive well in scavenging conditions, eat fallen grains in paddy fields, insects, snails, earthworms, small fishes and other aquatic materials. Ducks are hardy and can easily be adapted to various agro climatic conditions and they are likewise moderately resistant to diseases.

The population of poultry bird is 851.81 million as per the 20th livestock census. There is a tremendous growth (17.48%) observed in poultry farming from the 19th lives stock census from 729.21 million poultry birds to 851.81 million (20th livestock census) in India. The duck population increased from 2.35 million (19th livestock census) to 3.35 million (20th livestock census) and growth rate is around 42.36%, highest among the poultry sector. Duck occupies second place next to chicken in the production of table eggs in the country (Livestock Census, 2019). Highly duck populated states are West Bengal, Assam, Kerala, Bihar, Andhra Pradesh, Orissa and Tamil Nadu in India. In the North-East region, Assam, Manipur and Tripura possess 1st, 2nd and 3rd position, respectively in the duck population (Livestock census, 2019). There is a high chance of improving the productivity of ducks in India and duck production can provide self-employment for landless and marginal farmers.

There are all about total ten nos. of indigenous duck genetic germplasm available throughout the country but ICAR-National Bureau of Animal Genetic Resources, Karnal decleared Pati hans and Maithai as registered duck breed of India. Different indigenous and exotic duck genetic germplasm available in India is elaborately discussed below.

INDIGENOUS DUCK GENETIC RESOURCES OF INDIA

Sl. No	Indigenous Duck	Region
1.	Pati Hans	Tripura, Assam, Bangladesh
2.	Maithili	Motihari, Sitamarhi, Madhubani, Araria, Kishanganj and Katihar districts of Bihar
3.	Nageswari	Tripura, Assam, Bangladesh
4.	Tripura local	Tripura
5.	Manipur Local Duck	Different villages of Manipur
6.	Sada Pati	West Bengal
7.	Oddisa Local Duck	Different villages of oddisa
8.	Chara-Chemballi	Kerela
9.	Kuttanad	Kerala
10.	Sanyasi and Keeri:	Tamil Nadu

1.2 INDIGENOUS DUCK RESOURCES

1. Pati

Breeding Tract: Breeding tract of Pati hans is Brahmaputra and Barak valley of Assam, Tripura and neighbouring Bangladesh.

Characteristic Features

- Plumage colour of drake is dark brown with greyish black head and tail with black and white feathers
- Plumage colour of duck is solid brown
- In both sexes, white ring may or may not be present at neck region
- Color of bill, shank and feet are predominantly yellow
- Average body weight of adult bird is 1.58 kg.
- Male and female ratio in a flock ranges from 1:5 to 1:6

Production: Duck is mainly reared for meat and egg purpose. The age of first egg (AFE) is about 200 to 240 days. The annual egg production is ranges from 60 to 95 eggs.

2. Maithili

Breeding Tract: Maithili ducks are mainly distributed in Motihari, Sitamarhi, Madhubani, Araria, Kishanganj and Katihar districts of Bihar.

Characteristic Features

- These ducks have uniform light/dark brown feathers throughout the body with circular spots on the feathers (Mosaic pattern) in ducks.
- Drakes have dark brown to ash colour feather throughout the body.

- Head colour of drake is bright black to greenish black and brown in ducks.
- Body carriage is slightly upright and bill shape is horizontal.

Production Performance: Average age at first egg (AFE) is 191.12 days (range 159-223). Average annual egg production is 54.6 nos. (range 33-71). Average egg weight is 49.53g. Body weight at 6 month of age is 1.18kg (range 1.12-1.24).Population size is approximately 46,000.

3. Nageswari

Breeding Tract: Nageswari ducks are found in Tripura, Mizoram, Meghalaya, few pockets of Cachar and Karimganj districts of Barak valley of Assam and neighbouring Bangladesh. The original homeland supposes to be the erstwhile Sylhet district of Assam, now under the Bangladesh.

Characteristic Features:

- Nageswari duck is named because of their characteristic head-high snake like posture with a white stripe in the neck extending up to the breast along with bluish tinged eggs. It is also known as White Breasted Nageswari.
- Plumage colour is black or pencilled black.
- Mean adult body weights of male and female Nageswari ducks is 1.233 and 1.210 kg, respectively.

Production: They are egg type duck. The age of first egg (AFE) is average 188 days, ranges from 174-198 days. The average annual production of egg in back yard system is ranges from 100 to 150 numbers, while in proper management system can be ranged 200-220 nos. The average egg weight is of 60-69 g. The hatchability percentage in natural hatching ranged from 71.42 to 86.66% on total egg set.

4. Tripura Local Duck

Breeding Tract: They are the indigenous duck of Tripura, found in different villages of Tripura.

Characteristic Features:

- Plumage colour is dark brown.
- Body colour is covered by mostly mixed colour feathers. Head colour varied from green, black, white, brown, grey, grayish black and yellowish brown.
- Neck colour is black, white, brown, grey and yellowish brownish, black and white, with or without white ring.
- Bill, shank feet has predominantly yellow, other colour is orange.
- Mean adult body weights of male and female Tripura Local ducks is 1.24 kg. and 1.23 kg, respectively.

5. Manipur Local Duck

Breeding Tract: They are the indigenous duck of Manipur, found in different villages of that state.

Characteristic Features:

- Body colour is completely white, grey, greyish black, brown, white and grey.
- Head colour is green, black, white, blackish green, black and white.
- Neck colour is white and brown.
- Feather of tail is black and white.
- Bills are yellow, greenish yellow, blackish brown.
- Leg and feet colour is orange and blackish brown.

- Mean adult body weight of male and female Manipur local ducks is 1.757 and 1.689 kg, respectively.
- Average egg weight is 66.33 gm.

6. Sada Pati:

Breeding Tract: This duck is mainly distributed in the West Bengal.

Characteristic Features:

- Common duck of West Bengal is known as sada pati. Sada means "White".
- Feather colour is predominantly white.
- Body carriage is horizontal type.
- Shank is orange in colour. Bill is orange in colour with white colour bean.
- Drake has distinct male feather near the back curving upwardly.
- Ducklings are having yellow down feather during the birth which changed to white with maturity.
- Skin colour is yellow to white.
- Body weight of male ranges from 1305-1456gm and female has 1225 -1386gm.

Production: Hans lay an average of 130 eggs annually. The average age at first laying is around 135.5 ± 18.5 days.

7. Oddisa Local Duck

Breeding Tract: They are distributed in the villages of oddisa.

Characteristic Features:

- Local ducks have multiple colour plumage.
- Head colour in majority of ducks are the mixture of fawn

and white, other colour includes black, white and black-white spotted head.

- Neck colour of female is predominantly white, other colour includes chocolate colour.
- In duck the prominent bill colour is yellow followed by black and green.
- The drakes also have multiple colour plumage, majority of the colour is greenish black, and other colour includes brown.
- Neck colour in drake is predominantly white followed by brown colour.
- Few drakes have greenish colour neck and brown colour neck with white ring.
- Bill colour is predominantly yellow followed by yellowish green and greenish black with yellowish tint in drake.
- In both of the sexes, the body carriage is slightly upright, bill is horizontal. The breast colour is predominantly white, while wing colour is brown/ grey and white. The shank colour is yellow and orange, some with reddish colour. Eye colour is varied in both sexes. Skin colour is white.
- Egg colour of duck is cream in colour. The flock size small and varied from 9 to 30.
- The drake to duck ratio was usually 1:7. The adult body weight ranges from 1.32–1.53 kg.

8. Chara-Chemballi

Breeding Tract: These two types of ducks are native of Kerala state

Production: These ducks are reared mainly for egg purpose. Average age of flock at first egg (days) is 138-150

days. Ducks lay 150-200 numbers of eggs annually. The average egg weight is 61- 71.6 gm.

9. Kuttanad:

Breeding Tract: It is found in the Kerala.

Characteristic Features:

It is a variety of Chara and Chemballi.

10. Sanyasi and Keeri:

Breeding Tract: This variety mainly found in the area of Tamil Nadu. They are mainly reared for meat and egg purpose.

Sanyasi:

Characteristic Features:

- The Sanyasi female ducks are having saffron coloured plumage. White ring like feathers is present inconsistently in the neck within the variety.
- Bill colour of female is orange.
- Males are having dark brown plumage mixed with black. Head and neck covered with lustrous brown plumage.
- Bill colour is yellowish orange in males.
- Shank colour is orange for both males and females.

Keeri:

Characteristic Features:

- The Keeri female ducks are having mixture of black and brown plumage.
- Plumage of this duck has characteristic striations. Neck contain inconsistently white ring like feathers around the neck.

- In males plumage colour is a mixture of dark black and white colour.
- Head and neck is covered with lustrous black plumage.
- Bill and shank colour of females is grey / orange.

1.3 EXOTIC DUCK GENETIC RESOURCES AVAILABLE IN INDIA

Generally there are three categories of exotic breeds available in India.

A. Egg-type	B. Meat-type	C. Ornamental-type
1. Khaki Campbell	1. White Pekin	1. Crested White
2. Indian Runner	2. Muscovy	
	3. Aylesbury	

A. Egg Type

1. Khaki Campbell:

Location: It is an exotic duck breed reared mainly for egg purpose in India. This breed is developed by crossing between Mallard, Rouen and Runner ducks at Uley, in Gloucestershire, England by Mrs. A. Campbell. It is a prolific layer.

Characteristic Features: Generally, Khaki Campbell duck has three colour varieties i.e. khaki, dark and white. In both sexes, entire body is covered with khaki-coloured plumage. In Khaki Campbell drake, head is dark in colour usually olive green without white ring of its Mallard ancestors. The ducks are devoid of broodiness and sexual maturity reached at approximately 7 months. The weight of duck and drakes varies from 2-2.2kg and 2.2-2.4kg, respectively.

Production: The Campbell breed is an efficient layer and can lay an average of 300 eggs annually. Egg weight varies from 65 to 75 g and incubation period is 28 days.

2. Indian Runner:

Location: Indian Runners are mainly found in the Lombok, Java and Bali islands of Indonesian .

Characteristic Features: Indian Runner has mainly three varieties i.e.White, White Pencilled and Fawn. The body of Indian Runner duck is long, slim and round in appearance. The body carriage of this breed is upright. Pelvic girdle of Indian runner duck is located near to the tail region of the bird in comparison of other breeds of domestic duck which gives erect body carriage and allows to walk or “quickstep”, rather than waddle.

Generally, ducks has brown to white coloured feather. Head is long, wedge-shaped and shallow and neck is long, slender. Eyes are dark in colour. Bill colour is slate grey in ducks and yellow in drakes. Bean is absent in this bird. Appearance of this bird is racy with white skin colour.

A small curl is present on the tip of the tails of Drakes. Shank colour is orange. In ducks, height varies from 50 cm (20 in) while in drakes about 76 cm (30 in). In India, body weight of duck and drake ranges from 1150g and 1295g respectively, in scavenging system of rearing and under good managemental conditions body weight ranges from 1.5 to 2.6kg. Brooding ability of this breed is poor. Eggs are greenish-white in colour.

Production: The average age at first egg lay is 170 days and egg production of the ducks ranges between 150 and 280 eggs annually, average around 180 eggs. The average clutch size is 7–12 eggs.

B. Meat Type

1. White Pekin Ducks:

Location: It is a meat producing duck breed imported from Vietnam.

Characteristic Features: It has creamy white Plumage. Body is long, broad and deep. Bill and shanks are deep orange colour. Skin is yellow colour. It is a fast growing duck. It has good FCR, 1:2.3 to 1:3. The body weights of male and female ducks are 2.2 to 2.5kg in 42 days.

Production: It generally produces 150-160 eggs.

2. Muscovy Ducks

Location: It is originated in South America. In India, this breed is mainly found in West Bengal, Assam and other North eastern regions.

Characteristic Features: Muscovy duck is a true nesting duck. It has black and white colour plumage while sepia colour been rarely observed. Sepia coloured Muscovy are slightly smaller in comparison of black and white coloured Muscovy. They are commonly known as Chine hans or Marie hans. The ducks have a horizontal carriage. The caruncles are well developed in the drakes in compared to the ducks. The caruncles are of red in colour, extend from the bill to the eyes. Rudimentary caruncles are present in Sepia coloured Muscovy duck. The face has curly feather. The colour of eyes is yellow. The bills of the ducks are uniform in shape, slate grey in colour with white beans. Shank colour was completely slate coloured with some yellow markings over it. Skin is red in colour. Sepia coloured muscovy has yellow colour of shank and beak. Shanks have flippers. The skin colour is white. They have characteristic sexual dimorphism. The drakes having significantly higher body weight in comparison with the ducks. Drake does not have curly tail feather. The sexual maturity of this ducks range from 240 to 310 days. The body weight of drakes and ducks are 1950 to 2650g and 1332 to 1770g, respectively. They have broody nature. Ducklings are born with cream down with black markings while in sepia Muscovy Ducklings are khaki coloured with pink beak and yellow feet. These ducks prefer to stay at the banks of the water bodies rather swimming much. This ducks prefer foraging. These ducks are crossed with other breed of ducks resulted in development of Mule ducks which are sterile.

Production: They have slow growth rate and mature lately. Thus the age at sexual maturity is 348 days for the duck. They are very poor layers. The annual production of eggs ranges from 20 to 60, on an average around 28 eggs per year. The average clutch size is 5–6 eggs with long inter clutch interval. The average weight of egg ranges from 58-86 gm. The incubation period is 35 days.

3. Aylesbury

The colour of plumages is white. Legs are short and orange in colour. This breed is regarded as superior table bird due to its light bone and high percentage of creamy white flesh. Eggs are white in colour. This breed is excellent for meat and reaches market weight by 8 weeks.

C. Ornamental-Type

1. Crested White

Crested white breed of duck is known as an ornamental breed. Size is medium and used for dual purpose. Growth rate of this breed is high and also good layers. Adult body weight of drake is approximately 7 pounds and duck is 6 pound. Two varieties of crested white i.e. black and white but other varieties are developed by breeders such as grey, buff and blue.

CHAPTER 2

WHY DUCK FARMING?

DR. JOWEL DEBNATH

2.1 INTRODUCTION

Duck farming is a profitable livestock industry in India because of its egg, flesh and feather. Similar to chicken, ducks are mainly reared for eggs and meat. Egg size of ducks is comparatively larger as compare to chicken egg. Weight of egg is 4.5% of duck's body weight and weight of chicken egg is only about 3.3% of the hen's body weight. Furthermore, ducks are more productive than chicken and more adjustable to free-range system of rearing. Growth rate of duck is also higher than chicken. In comparison to chicken, ducks need simple housing.

2.2 SCENARIO OF DUCK FARMING IN INDIA

Ducks are mainly distributed in Eastern, North eastern and Southern states of the country and highly duck populated states are West Bengal, Assam, Kerala, Andhra Pradesh, Tamil Nadu, UP, Bihar and Orissa. In India, duck farming is concentrated in the hands of small and marginal and nomadic tribes. People of West Bengal and Kerala consuming more duck egg and meat in country.

2.3 ADVANTAGES OF DUCK REARING

1. Easy housing and easier to manage than chicken as compare to poultry.
2. Ducks are sturdier, could easily brood and are resistant to many avian diseases.
3. Ducks lay larger (15-20 grams) eggs and about 40-50 more eggs than chicken with higher nutritive value.
4. Ducks supplements their feed by foraging. They eat fallen grains in paddy fields, insects, snails, earthworms, small fishes and other aquatic materials.
5. Ducks work as exterminators of insects, snails and slugs etc.
6. Ducks requires lesser attention and thrive well in scavenging condition.
7. Duck do not require elaborate houses like chicken
8. Ducks lay more than 95% eggs in the morning prior 9.00 A.M. Thus saving lot of time and labour.
9. Economical rearing could be achieved with integrated farming process such as duck cum fish farming.
10. Duck small body feathers and the down are valuable use in industrial purposes.
11. Productive performances record of ducks i.e. age at first egg lay, eggs/duck/year, hatchability (%), body weight of mature ducks (kg) as 5-7 months, 70-190nos., 65-80% & 1.5-2 years, respectively. Production performances reflect more profit could be achieved by investing less time and labour.

CHAPTER 3

NUTRIENT REQUIREMENTS OF DUCKS AND THEIR FEEDING

DR. BIKAS CH. DEBNATH and TAPAN KR. DAS

3.1 INTRODUCTION

Duck occupies an important position next to chicken farming in India. As ducks are water fowl they are concentrated mainly in the rural areas of West Bengal, Assam, Andhra Pradesh, Orissa, Kerala, Tripura, Tamil Nadu, Bihar, Jammu & Kashmir and Manipur. Ducks are reared for eggs and meat purposes in India. Duck production has a great role in improving the socio-economical condition in rural India. Duck constitutes about 7-10% of the total poultry population of the world.

3.2 ADVANTAGE OF DUCK REARING

- They grow very fast. The feed conversion efficiency is 3-3.5.
- Laying birds are good layers yielding 250-300 eggs per year.
- Ducks are voracious eater of forages. So, feeding costs are reduced to some extent.

- Ducks are reared in pond and apart from compounded feed they eat snails, finger-lings, earth worm, insects and vegetation which also reduce the feed cost.

There are various breeds of ducks in India. Among them some of are good producers. Indian Runners, both white and fawn and white and Khaki Campbell are good breeds for egg production. They weigh between 2-2.5 kg.

3.3 KHAKI CAMPBELL DUCKS

Among the egg laying breeds, Khaki Campbell is the best producer. It is an egg type breed. They are originated by crossing a **Malaysian Indian Runner female (duck) with a Ruen male duck (drake).** They lay 250-300 eggs per year. Khaki Campbell ducks weigh about 2 to 2.2 Kg, and drakes 2.2 to 2.4 Kg. Egg size varies from 65 to 75 g. Khaki Campbell ducks, developed in Holland lay 335-340 eggs per duck in 365 days. The size of egg is 73.4 gram.

3.4 PEKIN DUCKS

It is originated from China. Of several breeds of meaty ducks, the Pekin ducks are found to the most popular meat type duck for table purpose in the world. **White Pekin ducks are broiler type**. In 8 weeks of rearing their body weight become 3.3-3.6 kg with a feed consumption of 9.6-9.9 kg. It is fast growing and has low feed consumption with fine quality of meat. It attains about 2.2 to 2.5 Kg of body weight in 42 days of age, with a feed conversion ratio of 1:2.3 to 2.7 Kg.

3.5 VIGOVA SUPER-M BROILER DUCKS

It is actually a cross between White Pekin duck and Aylesbury duck produced in Vietnam. Central Duck Breeding Farm

(CDBF), Hessarghata, Bangaluru imported this breed of duck from Vietnam. The ducks weigh 3 kg at 49 days of age.

The ducklings are reared for four weeks in the brooding house after hatching. During this time they should be properly fed with duck starter ration. The ration should contain adequate energy and protein level (2500 Kcal/kg and 20% protein) for their rapid growth. Ducks consume 10-12 kg feed in 8 weeks of rearing period. An adult female duck consume **170-230 g per day**. Ducks can be raised on dry or weight mashes. It is better to feed green feeds which can be chaffed and are of good quality like berseem, lucerne, etc.

3.6 PECULIAR ANATOMICAL STRUCTURE OF ALIMENTARY TRACT

Ducks **do not have a crop** and their **proventriculus is cylindrical**. Due to absence of a crop the ingesta passes very quickly which is actually quicker than the broilers. Due to peculiar structure of their bills it becomes difficult to take submerged food material as well as consumption of the most dry mash food particles of appropriate size. Most mashes form a sticky paste when mixed with saliva and *adhere to the papillae and other structures bordering the outer margin of the tongue and upper and lower bill.* This caking interferes with the movement of the food mass to the tongue resulting in reduction of food intake and increase in feed wastage which occurs when the duck attempts to shake or wash off the mash adhering to its mouthparts.

3.7 DRY MASH, WET MASH AND PELLET FEEDING

Duck mash can be compounded with normal feed ingredients used for poultry. The mash can be fed as dry or wet mash. Ducks have difficulty in swallowing dry mash, when they fed dry mash, they will take mouthful and swill it down in the nearest water source leading to heavy loss of nutrients. It is actually due to peculiar structure of the bill. To prevent this, wet mash may be given. Dry mash, crumbles and pellets can be given at free choice but wet mashes need to be given at frequent intervals. Generally wet mashes are given 4-5 times daily for ducklings up to 2 weeks of age and 3-4 times daily afterwards. After each meal residual wet mash is to be removed.

Although ducks have a preference for wet mash, nowadays pellet feeding has become popular in developed countries. Pellet size for duckling is 3.97 mm diameter and 8 mm length. After 2 weeks of age the size can be 4.76 diameters with 13 mm length. Pellet feeding in ducklings resulted in highest average weight both at 4 and 8 weeks. In the next ranks are, dry mash *ad lib* and 3 times wet mash feeding; dry mash *ad lib*; wet mash *ad lib*; and lastly 4 times wet mash feeding.

Pellet feeding also resulted in better feathering and general development than mash feeding at 12 weeks age. There was an outstanding advantage in egg production in the first and second laying year due to pellet feeding but mash feeding was superior in the third and fourth laying season. The net advantage was 2.1%. The same mineral mixture as recommended for poultry may be incorporated.

3.8 FEEDING BEHAVIOR

Ducks are voracious eaters and foragers too. Ducks may be grown on dry mash, a combination of dry and wet mash or pellets. Ducks prefer wet mash due to difficulties in swallowing dry mash. The pellet feeding, though slightly costly, has distinct advantages such as saving in amount of feed, minimum wastages, saving in labor, convenience and improvement in sanitary conditions. Ducks are good foragers. The use of range, pond or supplementary green feed, reduces the feed cost.

Ducks should never have access to feed without water. During the first eight weeks, birds should always have access to feed, but later-on they may be fed twice a day i.e. first in the morning and then late afternoon. Khaki Campbell duck consumes about 12.5 Kgs. of feed up to 20 weeks of age. Afterwards the consumption varies from 120 g and above per bird per day and depending upon the rate of production and availability of greens.

3.9 ROLE OF VITAMINS IN DUCK NUTRITION

Thiamin: Its deficiency causes paralysis of neck, leg and wing muscles. **Niacin:** Ducks are more susceptible to **bowed leg condition** and leg weakness resulting in complete crippling associated with niacin deficiency. Other general symptoms are lack of growth, diarrhoea and weakness. It is advisable to add niacin in the diet of duck ration. Brewer's yeast is the best source of niacin. **Vitamin C**: Vitamin C at the dose rate of 240 mg per liter drinking water is able to reduce the mortality of ducklings and also increased weight gain significantly. **Riboflavin:** In Riboflavin deficiency, the birds fail to grow after 2 or 3 days and usually die within 4-7 days. There appears to be an excess of secretion from the eyes and eyelids may become stuck together. There is a slow growth. There is no report of curled to paralysis. **Pantothenic acid**: Its deficiency symptoms are similar to riboflavin deficiency. **Pyridoxine**: Severe acute deficiency of this vitamin in ducklings causes failure of growth and severe anemia. Neither convulsion nor paralysis is observed in ducklings. A chronic deficiency in the older ducks causes lack of growth, paralysis, convulsions, severe microcytic anemia and poor feathering. **Biotin**: A deficiency leads to very poor growth. **Choline**: Its deficiency leads to **perosis (slipped tendon)**. **Vitamin A:** Its deficiency leads to poor hatchability and heavy early mortality, nasal discharge, paralysis. **Vitamin D3:** A deficiency of this vitamin causes rickets in ducklings, soft rubbery bones, rough feathering, poor growth, muscle degeneration, weakness. **Vitamin E**: This vitamin prevents **myopathy of gizzard and skeletal muscles.**

3.10 ROLE OF MINERALS IN DUCK NUTRITION

Copper supplementation results in increased growth rate and smaller caeca. A deficiency of **selenium** causes muscular dystrophy in ducks. Supplementation of **iodine** in the form of iodine causes increased growth rate, feed efficiency, and feathering and reduced carcass fat in ducklings.

3.11 ENERGY AND PROTEIN REQUIREMENTS OF STARTER AND GROWER AND LAYER DUCKS

Like chicken, ducks also eat primarily for energy and their feed intake is related to the dietary metabolizable energy levels. Ducks can meet their energy requirements if fed diets containing widely varying metabolizable energy levels provided that protein level of diet is increased with the increasing level. On higher energy diet the carcass fat also was reported to be higher. A decreased growth rate is also noted in duckling kept on low energy diet.

3.12 FEEDING OF STARTER DUCKLINGS

Immediately after hatching, the ducklings are reared for four weeks in the brooding house. During this time they should be properly fed with duck starter ration. A starter diet containing 20.5% crude protein and 2800 kcal ME/kg diet should be provided to the newly hatched duckling within 24-36 hours. The feed should have 1% calcium and 0.42% available phosphorus with all trace minerals and vitamins. There should have adequate riboflavin and niacin vitamin in the starter feed. Because, in riboflavin deficiency, the birds fail to grow after 2 or 3 days and usually die within 4-7 days whereas in niacin deficiency, duckling suffer from bowed leg condition and leg weakness resulting in complete crippling. Common salt should not be more than 0.30%. A high percentage of salt will cause watery droppings and wet litter condition. The mash food if offered to the ducklings then it should be moistened with water. Fresh drinking water must be provided to the ducklings round the clock.

3.13 FEEDING OF REARER DUCKS

A rearer diet containing 15% crude protein and 2500 kcal ME/kg diet should be provided to the grower ducks from 8 and it should continue up to 18 to 20 weeks. The feed should have 1% calcium and 0.35% available phosphorus with all trace minerals and vitamins. Excess feeding of growers during this period is not desirable as they will become fatty which will ultimately affect their reproductive performance. The growers should be slow growing during this period. Hence, a low protein diet is provided. Feed restriction is another option to slow down their growth. Early maturity is not advisable.

3.14 FEEDING OF LAYER DUCKS

A layer diet containing 18% crude protein and 2650 kcal ME/kg diet should be provided to the layer ducks from 20 weeks onwards it should continue till the cessation of laying. The feed should have 3% calcium and 0.35% available phosphorus with all trace minerals and vitamins. Around 5-6% oyster shell or limestone powder is need to be added in a layer ration in order to get the desired level of Ca% in the layer ration. The average daily feed requirement for a layer duck is 150g.

3.15 WATERING OF DUCKS

Clean drinking water should always be provided to the ducks. In some farming system, the feed and water is restricted during night hours in order to keep the litter dry. However, during hot weathers, when evening temperature is above 32°C then drinking water must be provided till the temperature of the shed decreased to 26.5°C.

Mineral mixture per 100 kg diet: Ferrous sulphate: 20 g; Manganese sulphate, 50 g; Zinc sulphate, 25 g; Copper sulphate, 1.5 g and Potassium iodate, 100 mg.

Vitamin mixture per 100 kg diet: Vit A, 800,000 IU; Vit. D3, 1, 00,000 ICU; Riboflavin, 400 mg; folic acid, 100 mg and Niacin 5 g.

3.16 AFLATOXICOSIS: A MAJOR PROBLEM OF DUCK REARING

Aflatoxins were first discovered as toxic metabolites of a fungus named *Aspergillus flavus* when huge mortality occurred among ducklings in early 1960 in England. Mycotoxins were found in moldy groundnut meal that was imported to England from Brazil. Mycotoxicosis is a condition caused by aflatoxin produced by the mould *Aspergillus flavus* in the feedstuffs such as groundnut, maize, sorghum, rice polish and other tropical feeds on storage. It is a great problem to the duck farmers particularly in tropical countries with changing agricultural technology. Improper drying of grains,

SUGGESTED NUTRIENT REQUIREMENTS FOR DUCK RECOMMENDED BY THE ICAR'2013

Characteristic	**Starter Duck (0-8 wks)**	**Grower Duck (8-16 wks)**	**Rearer Duck (16-20 wks)**	**Layer Duck (20 wks onward)**
Moisture, % (Max.)	11.00	11.00	11.00	11.00
Crude Protein, % (Min.)	**20.50**	**16.50**	**15.00**	**16.50**
Crude fibre, % (Max.)	7.00	8.00	8.00	6.00
Acid insoluble ash, % (Max.)	4.00	4.00	4.00	3.00
Salt, % (Max.)	0.30	0.30	0.30	0.30
Calcium, % (Min.)	1.00	1.00	1.00	3.00
Phosphorous (Available), % (Min.)	0.42	0.35	0.35	0.35
Linoleic Acid, % (Min.)				
Lysine, % (Min.)	1.00	1.00	0.08	1.00
Methionine, % (Min.)	1.00	0.75	0.60	0.75
Meth. + cystine, %	0.45	0.35	0.30	0.30
Metabolizable Energy (Kcal/Kg)	0.85	0.65	0.60	0.75
Min.	**28.00**	**2650**	**2500**	**2650**
Minerals and Vitamins:				
Sodium	0.17	0.15	0.15	0.17
Chlorine	0.12	0.12	0.12	0.12
Manganese, mg/kg	60.00	50.00	40.00	50.00

[Table Contd.

Contd. Table]

Characteristic	**Starter Duck (0-8 wks)**	**Grower Duck (8-16 wks)**	**Rearer Duck (16-20 wks)**	**Layer Duck (20 wks onward)**
Vitamin A, IU/Kg	3200	2250	2250	4000
Vitamin D3 , IU/Kg	400	350	350	650
Riboflavin, mg/kg	5.00	4.00	4.00	6.00
Pantothenic acid, mg/kg	10.00	8.00	8.00	12.00
Nicotinic Acid, mg/kg	60.00	55.00	50.00	50.00
Biotin, mg/kg	0.10	0.10	0.10	0.10
Folic Acid, mg/kg	0.60	0.40	0.40	0.60
Choline, mg/kg	1000	750	500	750
Vitamin E, mg/kg	20.00	20.00	15.00	20.00
Vitamin K, mg/kg	2.5	2.00	2.00	2.5
Pyridoxine, mg/kg	3.00	2.50	2.50	3.00

SUGGESTED NUTRIENT REQUIREMENTS FOR EGG AND MEAT TYPE DUCK USED AT CPDO, HESSARGHATA.

Characteristic	Starter Duck	Grower Duck	Layer Duck	Broiler(Vigova-M) Starter Duck	Broiler(Vigova-M) Finisher Duck
Moisture, % (Max.)	11.00	11.00	11.00	11.00	11.00
Crude Protein, % (Min.)	20.00	16.00	18.00	23.00	20.00
Crude fibre, % (Max.)	7.00	8.00	8.00	6.00	6.00
Acid insoluble ash, % (Max.)	4.00	4.00	4.00	3.00	3.00
Salt, % (Max.)	0.60	0.60	0.60	0.60	0.60
Calcium, % (Min.)	1.00	1.00	3.00	1.20	1.20
Phosphorous (Available), % (Min.)	0.50	0.50	0.50	0.50	0.50
Linoleic Acid, % (Min.)	1.00	1.00	1.00	1.00	1.00
Lysine, % (Min.)	0.90	0.60	0.65	1.20	1.00
Methionine, % (Min.)	0.30	0.25	0.30	0.50	0.35
Meth. + cystine, %	0.60	0.50	0.55	0.90	0.70
Metabolizable Energy (Kcal/Kg) Min.	2600	2500	2600	2800	2900
Minerals and Vitamins:					
Manganese, mg/kg	90.00	50.00	55.00	90.00	90.00
Iodine, mg/kg	1.00	1.00	1.00	1.00	1.00
Iron, mg/kg	120.00	90.00	75.00	120.00	120.00

[Table Contd.

Contd. Table]

Characteristic	Starter Duck	Grower Duck	Layer Duck	Broiler(Vigova-M) Starter Duck	Broiler(Vigova-M) Finisher Duck
Zinc, mg/kg	60.00	50.00	75.00	60.00	60.00
Copper, mg/kg	12.00	9.00	9.00	12.00	120.00
Vitamin A, IU/Kg	6000	6000	6000	6000	6000
Vitamin D3 , IU/Kg	600	600	1200	600	600
Thiamin, mg/kg	5.00	3.00	3.00	5.00	5.00
Riboflavin, mg/kg	6.00	5.00	5.00	6.00	6.00
Pantothenic acid, mg/kg	15.00	15.00	15.00	15.00	15.00
Nicotinic Acid, mg/kg	70.00	60.00	60.00	70.00	70.00
Biotin, mg/kg	0.20	0.15	0.15	0.20	0.20
Vitamin B12, mg/kg	0.015	0.010	0.010	0.015	0.015
Folic Acid, mg/kg	1.00	0.50	0.50	1.00	1.00
Choline, mg/kg	1300	900	800	1400	1000
Vitamin E, mg/kg	15.00	10.00	1.00	15.00	15.00
Vitamin k, mg/kg	1.00	1.00	10.00	1.00	1.00
Pyridoxine, mg/kg	5.00	5.00	5.00	5.00	5.00

rain and humid weather favor the mold growth. Among livestock animals ducks are the most susceptible to aflatoxins but the toxicosis is more harmful to the ducklings than the adult ducks. The ducklings particularly of Khaki Campbell are the most sensitive species to aflatoxin followed by Minikos and White Pekins. Day-old ducklings were identified as the species for the biological assay of aflatoxin. The LD 50 is as low as 0.3 mg/kg body wt.

Though chickens are also susceptible, the extent of toxicosis is less. Ducks can tolerate to the extent of 0.03ppm as against 0.2ppm in chickens. Ducks are about 200 times more sensitive than broilers and layers. The molds are not inherently toxigenic but can grow rapidly on the grains, stored feedstuffs and the metabolites that are liberated from the molds are very much toxic. They grow and colonize themselves rapidly in favorable environmental conditions like aerobic conditions with moisture content of feedstuffs exceeding 15% and an optimum temperature of 24- 25°C.

Among six types of naturally occurring aflatoxins, four major aflatoxins are B_1, B_2, and $G_1$and G_2. They are so named depending upon the color of fluorescence produced on the thin layer chromatograph plates under long wave UV light. $B_1$and B_2 fluoresces blue while $G_1$and G_2 fluoresce green color. The most important of these toxins is B_1 aflatoxin because of its toxicity and concentration in moldy feeds. The exceptional toxicity of aflatoxin B_1 for ducks is that this species contain a very high level of enzyme in the liver to convert aflatoxin B_1 to "aflatoxacol".

It has been noted that if feedstuffs are stored under similar conditions like same moisture and temperature, some feeds support the growth of aflatoxin producing molds while others not. The molds are present in the pure culture of three major feedstuffs like maize, groundnut cake and cottonseed cake which are highly potential for invasion by Aspergillus on storage. The main reason behind it is that this Aspergillus can produce toxin when present only in a pure culture in the feedstuffs and never produce toxin when contaminated with other fungi. Other feedstuffs which are less associated with aflatoxins are wheat, barley, oats, rice etc.

Principle target organ in aflatoxicosis is liver. Aflatoxin metabolism is 90 times faster in the duck liver than in rat liver. Aflatoxin B_1 is the most potent active hepatocarcinogen. The metabolites of B_1 bind with cell macromolecules such as DNA causing hepatotoxicity and carcinogenicity as a result of which necrosis and tumors are noted in liver. The toxins interfere with protein synthesis by the liver and ultimately synthesis of clotting proteins (prothrombins) is also inhibited. So, hemorrhagic lesions are seen on serous membrane and muscles. Even though aflatoxin cause serious damage to both growing ducklings and laying ducks, no aflatoxins have been found in eggs. No testicular damage is also found in case of drakes. Exposure to aflatoxins could reduce efficacy of vaccination due to immunosuppressive effect and various disease may flare up in the flock. The mold causes folic acid deficiency as a result of which anemia is noticed.

In ducklings and duck the aflatoxin produces liver lesions and results in huge mortality when present in high concentration. So, there is sudden death or signs of anorexia, depression, dyspnoea, oculonasal discharge, coughing, epistaxis, bloody droppings, staggering gait, ataxia and paralysis. In white skinned ducks, purplish discoloration of shank and web and hemorrhages on serous membranes and muscles may be seen. In ducklings, there are convulsions often characterized by opisthotonus, spasm of the neck muscles which leads to rapid death with the legs stretched posteriorly.

In sub acute and chronic cases, lower doses of aflatoxin produce chronic effects such as lethargy, unthriftness, lameness, hepatitis and delayed death. There is gradual loss of weight, reduced feed efficiency, and rough hair coat, falling of feathers, hemorrhages seen in serous membranes and muscles, anemia, enlarged abdomen, jaundice, reduced egg production in layers due to pathological changes in ovaries.

On postmortem there are petechial and ecchymotic hemorrhages on serous membranes and muscles, hemorrhagic enteritis, massive hepatic necrosis in ducklings, pale and enlarged liver with fat deposition, enlarged spleen and pancreas in acute cases of aflatoxicosis. Hepatic fibrosis, grayish, cirrhotic, mottled, and nodular surfaced liver hepatic carcinoma and bile duct hyperplasia, ascites, pale and swollen kidney with petechial hemorrhage is common postmortem lesion in ducks with chronic cases.

There is no specific treatment for aflatoxicosis. When the source of aflatoxin is removed from the feed, ducks make rapid recovery. Lipotropic drugs or liver tonics may be given to sick ducks. In chronically poisoned ducks contaminated feed must be with held and at the same time an easily digestible, low fat but a diet containing high levels of vitamin A, a high quality adequate protein including methionine supplementation should be provided. Methionine, about 30- 40% above normal level helps to detoxify the aflatoxins faster by liver. Increased level of vitamin K may be required where there is hemorrhage.

Always avoid moldy feeds. Feeds should be checked for aflatoxin particularly during and after rainy season. The most urgent need is to improve storage facilities of grains and other food materials, their transport practices so that contamination by molds can be reduced to minimum. During formulation of duck ration care must be taken not to use highly susceptible feedstuffs like maize, groundnut cake etc. Maize can be replaced by rice, wheat or other similar grains and groundnut cake can be replaced by other plant origin proteins like soybean meal, til cake etc. Mycotoxin binding agents like hydrated sodium calcium alumino silicate (HSCAS) may be used @ 0.5-1% of the ration of ducks. Fungistats in feed, such as Gention violet (2.05%), or Propionic acid (0.5%) may be added. In commercial methods, ammoniation of feedstuffs destroys the toxin effectively. A glucomannan based adsorbent proved that aflatoxin infected feed can be neutralized efficiently. It improved the liver function, thereby enhancing duckling performance. No residues of aflatoxins could be found in the meat. Directly sunlight can also destroy 50% of aflatoxins. Last but not the least, the moisture content of the stored feedstuffs should not be more than 10%.

3.17 FEEDING SCHEDULE OF PEKIN DUCKS

Of several breeds of meaty ducks, the Pekin ducks are found to the most popular meat type duck for table purpose in the world. **White Pekin ducks are broiler type**. In 8 weeks of rearing their body weight become 3.3-3.6 kg with a feed consumption of 9.6-9.9 kg. It is fast growing and has low feed consumption with fine quality of meat. It attains about 2.2 to 2.5 Kgs of body weight in 42 days of age, with a feed conversion ratio of 1:2.3 to 2.7 Kgs.

APPROXIMATE BODY WEIGHTS AND FEED CONSUMPTION OF WHITE PEKIN DUCKS (MALE) UP TO 8 WEEKS OF AGE

Age (weeks)	Body wt. (g)	Avg. daily intake(g)	Weekly avg. Feed intake (g)	Cumulative feed intake (kg)
0	60	-	-	-
1	270	30	210	210
2	780	110	770	980

Age (weeks)	Body wt. (g)	Avg. daily intake(g)	Weekly avg. Feed intake (g)	Cumulative feed intake (kg)
3	1380	160	1120	2100
4	1960	185	1295	3395
5	2490	200	1400	4795
6	2960	230	1610	6405
7	3340	240	1680	8085
8	3610	240	1680	9765

NUTRIENT REQUIREMENTS OF WHITE PEKIN DUCKS AS PERCENTAGES OR UNITS PER KILOGRAM OF DIET (90 PERCENT DRY MATTER)

Nutrients **ME Kcal/kg**	**Unit**	**0-2 Weeks** **2,900**	**2-7 Weeks** **3,000**	**Breeding** **2,900**
Protein	%	22	16	15
Arginine	%	1.1	1.0	
Isoleucine	%	0.63	0.46	0.38
Leucine	%	1.26	0.91	0.76
Lysine	%	0.90	0.65	0.60
Methionine	%	0.40	0.30	0.27
Methionine + cystine	%	0.70	0.55	0.50
Tryptophan	%	0.23	0.17	0.14
Valine	%	0.78	0.56	0.47
Macro-minerals				
Calcium	%	0.65	0.60	2.75
Chloride	%	0.12	0.12	0.12
Magnesium	mg	500	500	500
Nonphytate phosphorus	%	0.40	0.30	
Sodium	%	0.15	0.15	0.15
Trace minerals				
Manganese	mg	50	-	-
Selenium	mg	0.20	-	-

[Table Contd.

Contd. Table]

Nutrients ME Kcal/kg	Unit	0-2 Weeks 2,900	2-7 Weeks 3,000	Breeding 2,900
Zinc	mg	60	-	-
Fat soluble vitamins				
A	IU	2,500	2,500	4,000
D3	IU	400	400	900
E	IU	10	10	10
K	mg	0.5	0.5	0.5
Water soluble vitamins				
Niacin	mg	55	55	55
Pantothenic acid	mg	11.0	11.0	11.0
Pyridoxine	mg	2.5	2.5	3.0
Riboflavin	mg	4.0	4.0	4.0

[b] Question marks indicate that no estimates are available.

3.18 FEED INGREDIENTS FOR DUCKS

Cereals and their by-products:

For the proper feeding of duck a certain portion of the ration includes cereal and their by-products. In the case of duck, cereals are the main component of their daily rations. In this regard, they compete directly with the human beings for cereal feeding. Cereal comes under *Graminae* family.

CEREALS:

Maize (*Zea mays*): In case of duck, the main source of energy is supplied through maize. It is the richest source of energy. Yellow maize contains a pigment, cryptoxanthene, which is a precursor of vitamin A. Maize is low in protein. Proteins present are of low quality. Maize is deficient in tryptophan, methionine and lysine. Maize contains 73% starch on dry matter basis, very low in crude fibre and has a high ME value

FEED FORMULAE FOR DUCKS FOLLOWED AT C.P.D.O. (SR)

INGREDIENTS (%)	KHAKI CAMPBELL DUCK			WHITE BROILER DUCK		
	Starter	Grower	Layer	Starter	Grower	Layer
WHEAT	45	48	42	60	40	40
YELLOW MAIZE	-	-	10	-	29	20
D.O.R.B.	14	25.5	6.5	-	10	-
SOYABEAN MEAL	25	15	20	25	10	20
FISH MEAL	10	6	10	10	6	10
LUCERN LEAF MEAL	2	2	2	2	2	2
MINERAL MIXTURE	2.5	2.5	2.5	2.5	2.5	2.5
SHELL GRIT	-	-	5.5	-	-	5
D.C.P.	1.0	0.5	1.0	-	-	-
VITAMIN MIXTURE	0.5	0.5	0.5	0.5	0.5	0.5
TOTAL	**100.0**	**100.0**	**100.0**	**100.0**	**100.0**	**100.0**
Vitamin mixture in g (per 100 kg)						
VITAMIN AB_2D_3 K	25	25	30	25	25	30
VITAMIN B+E	25	25	30	25	25	30
NIACINAMIDE	5	5	5	5	5	5
CHOLINE CHLORIDE	50	50	50	50	50	50
ANTIBIOTIC	50	50	50	50	50	50
A.P.F. – 100	20	20	20	20	20	20
U.T.T.P.	20	20	20	20	20	20
TOTAL	**195g**	**195g**	**205g**	**195g**	**195g**	**205g**

FEED SCALE FOR KHAKI CAMPBELL DUCK

Age (Weeks)	Feed consumption/ bird/week/Kg.	Age (Weeks)	Feed consumption/ bird/week/Kg.
1	0.115	13	0.595
2	0.255	14	0.605
3	0.425	15	0.630
4	0.620	16	0.705
Total	**1.415**	**Total**	**2.535**
5	0.720	**Progressive total**	**9.945**
6	0.770	17	0.615
7	0.785	18	0.655
8	0.790	19	0.665
Total	**3.065**	**20**	**0.745**
Progressive Total	**4.480**	**Total**	**2.680**
9	0.690	**Progressive Total**	**12.625**
10	0.730	21	0.775
11	0.755	22	0.945
12	0.755	23	0.950
Total	2.930	24	0.955
Progressive Total	**7.410**	**Total**	**3.625**
		Progressive Total	**16.250**

(3340 Kcal ME/kg). Maize is high in lenoleic acid, which is an important factor in diet controlling the egg size of ducks. Inclusion level in case of starter and grower and layer duck is up to 55 and 50%, respectively.

Wheat (*Triticum aestivum*): In wheat grain, crude protein ranges from 8-14% (on Dry Matter Basis). Wheat, especially if finely milled, forms a pasty mass in the mouth of ducks which leads to digestive upset. Its by-product, wheat bran contains 12-14% crude protein. Wheat contains more of B-complex vitamins than maize. Inclusion level in case of duck is up to 50%.

COMPUTATION OF DUCK RATIONS

Composition of Duck rations for three different phases

Ingredients	Duck starter (%)	Duck grower (%)	Duck layer (%)
Maize/broken wheat	54.3	60.3	57.3
Wheat Bran	14	19	9
Soybean meal	19.5	10.5	18
Dry fish	10	8	8
Bone meal	1	1	1
Mineral mixture	1	1	1
Oyster shell	-	-	5.5
Salt, max.	0.2	0.2	0.2
Toxin binder	0.08	0.08	0.08
Antibiotics/100kg	20g	20g	20g
Vitamin AB_2D_3 /100kg	20g	20g	20g

N.B. As maize has been used instead of wheat; the level of mycotoxin binder has been increased accordingly (from normal 0.05% to 0.08%).

DAILY FEED INTAKE AND GAIN IN BODY WEIGHT OF KHAKI CAMPBELL DUCK

Mash/ Pellet	Age (Weeks)	Avg. daily feed intake /duck(g)	Total feed intake/duck/ week(g)	Live weight/ duck(g)	Progressive feed intake
Duck starter (0-8 weeks)					
	0 day	-	A little	30 g	
	1st week	15	105	80g	
	2	35	245	120g	
	3	60	420	250g	
	4	90	630	**350g**	
		Total	**1400 gram**		**1400 gram**
	5	105	735		
	6	110	770		
	7	115	805		
	8	115	805		
		Total	**3115 gram**		**4515 gram**

[Table Contd.

Contd. Table]

Mash/ Pellet	Age (Weeks)	Avg. daily feed intake /duck(g)	Total feed intake/duck/ week(g)	Live weight/ duck(g)	Progressive feed intake
Duck grower (9-20 weeks)					
	9	100	700		
	10	105	735		
	11	110	770		
	12	110	770		**7490 gram**
	13	85	595		
	14	90	630		
	15	90	630		
	16	100	700		**10045 gram**
	17	90	630		
	18	95	665		
	19	95	665		
	20	105	735	**2000g**	**12740 gram**
		Total	**8225 gram**		
Duck layer (21st week onwards weeks)					
	21	110	770		
	22	135	945		
	23	135	945		
	24	140	980		**16380 gram**
		Total	**3640 gram**		

PERFORMANCE CHART OF KHAKHI CAMPBELL (EGG TYPE)

1	Age at first egg	120 days
2	Age at 50% production	146 days
3	Annual Egg Production	250 eggs
4	Egg weight at 40 weeks	66 g
5	Body weight at 40 weeks	1.80 kg

6	Daily feed consumption per bird	120 – 130 g
7	Ducklings mortality (0-8 weeks)	2 – 3%
8	Grower mortality (8 – 20 weeks)	0.2 – 0.5%
9	Adult mortality (20 – 72 weeks)	5 – 7%

PERFORMANCE CHART OF VIGOVA SUPER-M (MEAT TYPE)

1	Day old body weight	47 – 48 g
2	Body weight at 4 weeks	1.3 – 1.5 kg
3	Feed consumption up to 4 weeks	3.0 – 3.2 kg
4	Body weight at 6 weeks	2.3 – 2.5 kg
5	Feed consumption up to 6 weeks	5.8 – 6.2 kg
6	Mortality (0-6 weeks)	2 – 3%

AVERAGE BODY WEIGHT AND FEED CONSUMPTION OF BROILER DUCKS AT 6 WEEKS OF AGE

			Feed Consumption	
Age (Weeks)	Body Weight (Weeks)	Daily feed intake (g)	Weekly (Kgs.)	Cumulative (Kgs.)
1.	0.183	20	0.140	0.140
2.	0.526	80	0.560	0.700
3.	1.048	144	1.008	1.708
4.	1.533	180	1.260	2.968
5.	2.082	203	1.421	4.389
6.	2.498	200	1.400	5.789

Rice (*Oryzae sativa*): Rice is a good source of energy. It is highly palatable and digestible. It has a ME value of 2845 Kcal/kg. Inclusion level in both ducklings and duck is 0-40%.

Sorghum/ Jowar (*Sorghum vulgare*): Seed coat of sorghum is quite hard. It is difficult to digest also slightly less palatable to maize. It is used after crushing and grinding

DAILY FEED INTAKE OF VIGOVA SUPER–M BROILER DUCK IN RECBF, RK NAGAR, TRIPURA

Day	Feed intake (g)	
1st	2.3 g	96g/week i.e 13.7 g/day/duckling
2nd	7	
3rd	10.5	
4th	14	
5th	17	
6th	21	
7th	24.4	
8th	29.3	331g/week i.e. 47.2g/day/duckling
9th	36	
10	40	
11	45	
12	55.5	
13	60	
14	65.2	
15	70	580g/week i.e 82g/day/duck
16	74	
17	79.2	
18	83.3	
19	87	
20	91	
21	95	
22	95	754g/week i.e. 107g/day/duck
23	102	
24	105.5	
25	108.3	
26	112	
27	113.2	
28	117.2	
29 to 120 days	150-200	
Adult (120 days and above)	190-240	

for mixing with other ingredients. Protein content of this grain varies from 8.5-12.5%. ME value is 2645 Kcal/kg. Inclusion level in duckling is 0-10 % (dark) and 0-25 %(white variety). In adult ducks it is 0-20% and 0-40% in dark and white variety, respectively.

Bajra (*Pennesetum typhoides*): Feeding value of bajra is quite similar to that of sorghum. Crude protein content is 10-12%. This grain is also rich in tannins. It can be used in place of maize in poultry feeding. ME value is around 2650 Kcal/kg. In ducklings and adult ducks inclusion level is 0-25% and 0-45%, respectively.

CEREAL BY-PRODUCTS

Wheat Bran: Wheat bran is the coarse outer covering of the wheat kernel. Wheat bran averages 12.5% moisture, 13% in crude protein, 12% crude fibre and 4.5% in fat and in good quality brans should not contain more than10% crude fibre. The protein content of wheat bran is of higher quality than that of the entire wheat grain or maize. Wheat bran is rich in niacin and fairly high in thiamin, but low in riboflavin. It is also laxative in nature, owing to its crude fibre and pentosans contents. It is tasteful but digestibility is low. It has 0.07% Ca and 0.35% P. It is rich in phosphorus, though due to presence of phytic acid most of it is not available to the ducks. Its inclusion level in duck is 10-15%.

Rice Bran: Rice bran of good quality contains as much protein and nearly as much fat as rice polish, but averages 11-12% in crude fibre. It contains 12-14% crude protein and about 12% of oil. It contains 0.06% Ca and 1.12% P. It is rich in thiamin and very high in niacin and also rich in other B-complex vitamins. It is a rich source of manganese. It is because of high fat content, rice bran develop rancidity. It is used for feeding of ducks. Its inclusion level in duck is 10-20%. Nowadays, to avoid the rancidity problem of fat, de-oiled rice bran (DORB) is commonly used. The crude protein percentage is higher in DORB than rice bran.

Rice Polish: Rice polish of good quality averages 12.4% in crude protein and 13.6% in fat, with only 2.7% crude fibre. Good quality rice polishing is rich in energy. Rice polish supplies as much total digestible nutrients as maize. Its ME content is around 3300 Kcal/kg in duck. Its inclusion level in ducklings and adult ducks is 0-40% and 0-50%, respectively. They are excellent sources for many of water soluble vitamins particularly thiamin and niacin. It is because of high fat content, rice polish develop rancidity. It is because that rice polish contains unsaturated fatty acid. It is used for feeding of duck. Inclusion level in case of duck is 10-30%.

PLANT ORIGIN OIL CAKES

Groundnut Cake (GNC): GNC is most widely used, high protein feed. In expeller variety CP content is 45% and 10% fat. GNC is an excellent source of arginine but deficient in lysine, methionine and cystine. First limiting amino acid is lysine. Toxic factor present in the GNC is 'Aflatoxin' a metabolite of the fungus *Aspergillus flavus*, particularly in warm rainy season. The cake tends to become rancid especially in warm moist climate. It should not be stored more than 6 weeks in summer or 3-4 months in winter. Ducklings are highly susceptible. The content of oil is variable according to the process of extraction of oil. In ducks, inclusion level is 0-40%. In poultry, ME of GNC is around 2600 Kcal/kg, inclusion level in both chicks and adult poultry is 0-40%.

Soybean Meal (SBM): Normally solvent extracted. Oil content is 1%. Most of the SBM are deoiled type. It is an excellent source for lysine but methionine is the first limiting amino acid. Two grades of SBM: 44% CP and 49% CP. SBM is poor source vitamin B-complex, and must be provided either as a supplement or in the form of an animal protein such as fish meal. The oil in the soyabean has a laxative effect and may result in the production of soft body fat. In poultry ME value is around 2700 Kcal/kg and its inclusion level in both ducklings and adult ducks is 0-40%.

Deoiled Mustard Cake: Oil content is less (1-1.5%) in Deoiled Mustard Cake. Crude protein content is 39%. Ca and P content are very much higher (0.29% and 0.39%, respectively). The protein is deficient in lysine Deoiled type can be used for duck up to 10% of the ration. It contains a goitrogenic substance which reduces the growth rate in ducks. About 10-15% of this cake can be incorporated in duck ration.

Sunflower Meal: Good quality sunflower meal contains 40-44% high grade protein especially rich in methionine (in decorticated variety). The first limiting amino acid is lysine. Sunflower seed meal is a satisfactory substitute to GNC in starter duck ration and it can replace 100% GNC without any adverse effect on weight gain and feed efficiency. This cake is also safe for layers for replacement of GNC without any adverse effect on egg production and egg weight. Its ME value is 2230 Kcal/kg and inclusion level in duckling and adult duck is 0-10 and 0-20%, respectively.

Safflower Meal: In decorticated form it has about 40-45% crude protein, while the values go down to about 18-20% if not decorticated. Even decorticated type contains about 14-16% crude fibre. Safflower meal is low in lysine and methionine. Whenever safflower meal is fed to ducks it should be used in conjunction with other lysine rich protein concentrates (due to low lysine content).

Sesame Cake/Til Cake: It is a good source of protein, next to GNC. Solvent extracted cake has 47-50% crude protein. The protein in this cake is rich in leucine, arginine, tryptophan and methionine but is relatively low in glycine, cystine and lysine. The residual oil of this cake/ meal is highly unsaturated and may result in soft body if consumed in excessive amount. The cake is palatable with laxative action. Inclusion level in ducklings is 0-10%, and grower and layers it is 0-20%. Crude protein 40-50%, crude fibe 5%, ether extracts 7%, NFE 34.3%. Its ME value in poultry is 1882 Kcal/kg. Inclusion level in ducklings and duck is 0-10 and 0-20%, respectively.

ANIMAL ORIGIN PROTEIN SOURCES

Fish Meal: Protein content of various fish meal varies over a range of about 50-75% but the composition of the protein is relatively constant. It is rich in all essential amino acids, particularly lysine, cystine, methionine and tryptophan. A good quality fishmeal should have CP 55%, ether extract 6.9%, minerals 25%, Ca 7.9% and P 4.4%, 4.82% lysine and 1.52% methionine. They are good source of vitamin A, D and B-complex vitamins, particularly choline, patothenic acid, vitamin B_{12} and riboflavin. Fishmeal is the richest source of Vit B_{12}. Its ME value in case of poultry is around 1850 Kcal/kg and inclusion level for both ducklings and adult duck is 0-10%.

Meat Meal: Meat meal is rich in crude protein (50-55%) and ash (21%) with high calcium (8%) and phosphorus (4%). Low in trytophan and methionine (first limiting amino acid is tryptophan). The low methionine and tryptophan levels in this meal reduce their values since they cannot adequately make good the deficiencies of these amino acids in the high cereal diets of ducks, especially when proportions of maize are given (maize is very low in tryptophan). It is an important source of

Ca and P. It has ME value of 2000 Kcal/kg in poultry and inclusion level for both ducklings and adult duck is 0-5%.

Blood Meal: It has 80% crude protein, small amount of ash and oil and 10% moisture. It is poor in Ca and P (0.33% Ca, 0.26% P). This meal is one of the richest sources of lysine (6.9% lysine). Its inclusion level for ducks is 1-2%. Its ME value in case of poultry is around 2850 Kcal/kg.

Silk Worm Pupae Meal: A good quality silk worm pupae meal could be prepared containing 55% crude protein and 25-27% oil. After de-oiling, keeping quality of the meal is improved. It can replace whole of fishmeal in poultry rations. However, replacing 60% fish meal would give better results.

3.19 COMMONLY USED FEED ADDITIVES IN DUCK RATIONS

Antibiotic in ducks: The good effect of feeding antibiotic feed supplement is seen in ducks given all vegetable protein diets than those getting animal protein diets. The growth rate is better when antibiotic is used in an unhygienic condition than when used in hygienic condition. In old buildings there is increase of 10-15% in the growth rate of ducks. Due to antibiotic feeding the increase in growth rate may vary between 10-20% and reducing the feed intake by about 2-5%. Antibiotic improves the efficiency of feed utilization by 5-8%. The dose rate is 5-15mg/ kg feed and no advantage is seen in exceeding these levels. e.g. Virginamycin, Bacitracin, Flavomycin etc.

Probiotics: Probiotics are live microbial feed supplements which beneficially affect the host animal by improving its intestinal microbial balance. Overall performance of the duck is improved due to the use of probiotics in ducks. Microorganisms those are as probiotics: *Lactobacillus acidophilus, Aspergillus oryzae; Saccharomyces cerevisiae* etc.

LEVEL OF INCLUSION OF COMMON DUCK FEED INGREDIENTS

Feed ingredients	%
Maize	60
Sorghum	30-40
Bajra	10-20
Wheat	50
Rice	40
Rice bran	10-20
De-oiled rice bran	10-20
Rice polish	10-30
Wheat bran	10-15
Tapioca meal	5-15
Molasses	0-5
Maize gluten	0-10
GNC	10-30
Sunflower cake	10-20
Safflower cake	5-15
Mustard cake	0-5
Soybean meal	40
Fish meal	5-10
Meat meal	5-10
Blood meal	3
Silkworm-pupae meal	6

Fructo-Oligosaccharides (FOS) and Mannan-Oligosaccharides (MOS): The beneficial effects of yeasts are due to the cell wall component of the yeast. They are complex carbohydrates available in yeast cell wall. They improve health and performance. They inhibit (block) the attachment of pathogenic harmful bacteria to the intestine. Ultimately prevents colonization and disease is prevented. Proliferation of the beneficial (probiotic) bacteria takes place which inhibits the growth of more harmful bacteria.

Antioxidants: The addition of antioxidants mops up the free radicals. Antioxidants are of two types: natural and synthetic. Vitamin E and vitamin C are natural antioxidant while Butylated hydroxyanisole (BHA), Butylated hydroxytoluene (BHT) and Ethoxyquin are the most common synthetic antioxidants.

Mycotoxin Binder: Mycotoxins are produced from various fungus which are harmful to human health. Some important fungus are *Aspergillus, Fusarium, Penicellium* and important mycotoxins are aflatoxins, zearalenone, ochratoxin A, etc. Mineral clays like aluminosilicates (Hydrated sodium calcium aluminosilicate i.e. HSCAS), etc. are commonly used toxin binders.

Enzymes as Feed Additives: Maize and soyabean meal are the most preferred feeds for poultry including ducks. Due to high cost of these ingredients, feed stuffs like wheat, oats, barley, etc. and sunflower seed meal, rapeseed meal, etc are being used nowadays. But unfortunately, the digestibility of these ingredients is less. The presence of non-starch polysaccharides (NSPs) in the cereal grains are cell ulose, araboxylans and betaglucans. Hence, forcommercial enzymes like xylanase, arabinase, Mannase, glucanase, pectnase, cellulose, phytase etc. are used in the duck ration.

CHAPTER 4

DUCK MANAGEMENT

DR.JOWEL DEBNATH AND DR. DEBAJYOTI SARKAR

4.1 INTRODUCTION

Duck is a sturdy and prolific animal. Life span of duck ranged from 6 years to 10 years and productive life is up to 4-5 years. Duck has a high adaptability, disease resistant than any other poultry birds. Ducks are visible in the marshy, water logging area, supplement feed by foraging, less feed requirements than other poultry birds and require only house just for night halts. Although ducks are very smart and can be trained easily. Duck can easily recognise the place of foraging or watering or feeding and love to stay in the watering place for the day. When ducks are left in the morning, they easily move to the selected place and they majorly return to the home in the evening.

There are two major constrains for rearing of duck. First, diseases which usually appear in the certain season and may cause heavy mortality, secondly, predators which are usually reside within the area of house or the foraging area. The most problem comes with these birds with the health management. In Indian situation, farmers primarily rear ducks as a subsidiary part of animal rearing. So majorly they concern about the egg production and meat production. Farmers are very aware of the season at which ducks are mostly affected and that time either they use to go for sale or used for table purpose.

4.2 REARING SYSTEMS OF DUCK

1. **Traditional or Scavenging System or Herding System:** It is a completely natural system of rearing. Ducks are responsible for their own food. They are left out for the foraging in the morning and return back to house by their own.
2. **Semi-intensive System**: Ducks are moving within a confined area, 50% by natural foraging and rest feed is in the form of concentrate.
3. **Intensive**- Ducks are in confinement. They receive the food by the rearers.

Rearing systems of duck also can be categorised in following ways:

a. **Range System:** In this rearing system, ducks are only kept inside or enclosed at night and in the day the ducks are unobstructed to wander outside in search of feed. The ducks only require night shelter and nests for laying eggs.

b. **Restricted System:** The ducks are kept enclosed everlastingly in a confined shelter or with a run in the open. In this system of rearing, the ducks will reside in the same place. It is very convenient to keep watch without stress.

c. **Internal System:** In this system of rearing, large number of ducks is maintained with minimum labour. In this system needs more asset than other two systems of housing. Farmers have to provide all nourish and water and clean the shed in frequent way.

4.3 TRADITIONAL HOUSING SYSTEM

An ideal Duck houses should be well ventilated, dry, clean and above the ground level to protect them from the access of rat and predators. Ducks spend most of the time in water / outdoor compared to indoor and makes the body and feather wet. Such type of cleaning and dry sheltered help them to keep clean, peer their feather which makes their feather water proof, protect skin and make the body warm.

In villages, the farmer usually made their house by locally available materials. They generally never build permanent shelter house except few progressive farmers. The shelter houses are mainly constructed by using the bamboo and wood. Farmers kept the height of the shelter house at a height of 1.5 to 2 feet from the ground level that protects ducks from the harsh climate, predators. Farmers are with small flock size, generally kept under the bamboo made basket during the period of night.

Fig. Traditional Duck Rearing System

Housing for Ducklings

The ducklings are generally kept under supervision to prevent attack from the predator and reared under intensive system at early days.

Housing for Growers

The grower ducks are somewhat self resilient to save themselves from the attack of predator and under supervision. They are reared in semi-intensive system. They are guarded by fence of about 2ft height. The free space required for movement is 10 sq ft per duck.

Housing for Layer Ducks

Layer ducks are reared in semi-intensive system. Ducks are reared with sufficient area for movement 15 sq ft per duck. The male and female ratio is 1:6 to 8 for better fertility. Before leaving the ducks for foraging close observation should be for egg production. Although, duck lay egg in the morning time. The area of foraging should also be monitored for the eggs.

Housing for Brooders

The housing of rearing is almost similar to that of chicken. The brooder house has 32°C to 37°C temperature maintained. The temperature provided by the electric bulb. At this stage ducklings are unable to maintain their own body temperature. So, constant maintaining of temperature is highly desired.

Ventilation

Proper ventilation in a house is the vital requirement for animals. Although, duck houses or shelters with small flocks usually do not require any mechanical ventilation as farmers usually made the houses with locally available cheaper materials and always look after the matter of ventilation intentionally or unintentionally to reduce the cost of housing.

4.4 INTENSIVE AND SEMI INTENSIVE SYSTEM OF REARING

Housing

House generally made up of the concrete. Space requirement is at least 0.2 m^2 of floor space/duck means 5 birds/m^2. Housing should be combination of slatted area and litter area. Ducks can lay and take rest in the litter area and can walk in the slatted area. This type of mixed housing reduces the lameness in the ducks. There should be strict separation by means of barricade or wall for each age group of the ducks.

Ventilation

In intensive farming system, when large flock stays together in a room require proper ventilation. Lack of ventilation leads to accumulation of the noxious carbon dioxide and other gases produces from the litter. These gasses are directly affecting the respiratory system of the ducks. It also reduces the immunity level of the ducks leading to much outbreak of disease.

Windows or ridge ventilation can be adopted to maintain the cross air flow. In case of much congested flocks, fans can be used to exchange the inner air to outside. In modern duck buildings ventilation systems should deliver a minimum @0.2 cfm/lb duck weight at 0.05 inches (water gauge) static pressure. When temperature is above the permissive level, there should be an increase in the cross ventilation rate, a maximum ventilation rate @0.8 cfm/lb duck weight at 0.02 inches static pressure.

Waterers for Ducks

Water is the source of the many diseases for human as well as for the animals and birds. Ducks are spending much time in water, although they should get clean water which prevents them to get additional harmful bacteria and toxin into their gastrointestinal tract. The waterer are much more prone to get heavy loads of microorganism due to presence of saliva along with feed particles. Waterer should be cleaned every day or alternate day for reducing the bacterial loads.

Different waterers are available for ducks such as automatic trough, cup or plasson waterers, nipple waterers. To provide and for easy access to water, the water trough should have depth minimum at least 4 cm so that bill can be submerge properly into the trough and ducks can drink the water properly. In deep litter system, watrerer should be at the height of the duck to prevent spillage of water in the bedding materials. To prevent the same wire-mesh screen also can be used. Starter and grower ducks need a minimum about 1 inch (2.5 cm) of watering space per duck. For adult and laying breeders duck need a minimum about 2 inches (5.0 cm) per duck.

In case of nipple waterers need 15 nipples per 100 ducks for starter and growers and 20 nipples/100 ducks for adult and laying breeders. To use this kind of waterers duckling should get trained at early age of life.

Ducks rearing with adequate place can provide a pool outside run area. This concrete made pond should be 1 m wide by 0.25 m deep or saucer -shaped pools 2 m wide and 0.25 m deep with good drainage facility.

Management of Litter and Yards

Ducks are fond of water. They drink much water hence; excreta contain over 90% moisture compared to chickens. In intensive rearing system maintenance of the litter in dry condition is one of the major tasks for the rearers. This required regular addition of the fresh litter materials and cleaning of old dumped litter. While in semi intensive system, ducks spend most of their time in outdoors which leads to less spoilage of the litter materials. Although during the winter time water should be provided in to the house itself. The raking and changing in the litter materials in regular time interval is very much crucial for the health of the ducks. The depth of litter should be of about 7–8 cm.

The yard should be clean and free of any holes, through which any predators can reach to the ducks and affects the production and productivity of the ducks either by losing of ducks or eggs. The dirty yards also invite multiple infections to the ducks as they are the reservoirs of multiple type of organism. It should be little sloppy so that water should not get stagnated in any place.

Fig. Management under deep litter system

Lighting

Ducks are sensitive for the light. The production of egg is increased with the increase of the day length. During the laying period there should be good amount of the light in the house. Another way, if light is not provided the productivity of ducks will be constricted to a certain period of months in a year. To get eggs for prolong period of time need of provision of artificial light in addition of day light. The lightening period ranges from 14-17 hours in a day. Constant lighting for this much time periods prevent any decrease in egg production due to the continuous stimulation to hypothalamus reach via optic nerve. This helps to maintain continuous laying for 7-12 months.

The intensity of light is minimum 10 lux (1 foot candle), as the duck's eye at this level stimulating sexual response in both sexes. Although, during breeding and laying recommended level of light intensity is 20-30 lux. This minimum light intensity of 10 lux can be achieved by providing one 60 watt incandescent bulb for every 18 m^2 of floor space.

Ducklings required more artificial light in comparison to the growers to collect feed and water at night. Although at night, application of dim light is required to prevent feather picking by the ducks. This also prevents panicking within the flock following black out of light. To prevent the stress one 15 watt bulb for every 18 m^2 of floor space also can be used.

Grower ducks require constant period of light during the period of initiation of laying. Pekin ducks with provision of constant application of light for 17hrs/day yield maximum production. To maintain the constant lightening period, it has to remember that there will be constant increase or decrease length of the day by 15 minutes each week of each month of summer and winter. There should be constant monitoring of the light daily wise so light period can be maintained at constant period throughout the year.

Floor Space

Ducks require space for their own movement as well to take own care. Spacing also prevents overcrowding which has direct impact in their growth, health and production. The space recommendation has given below.

Floor Space allowances for ducks

S.I. No	Age in Days	Space/Duck (sq cm)	Space/Duck (sq ft)
1	1	289	0.31
2	2	576	0.62
3	3	1024	1.10
4	4	1369	1.47
5	5	1764	1.90
6	6	2116	2.28
7	7	2304	2.48
8	Developing breeders	2500	2.69
9	Laying breeders	2809	3.02

Source: https://www.vet.cornell.edu/

Flooring for Ducks

Flooring is a vital for sustainable duck farming. The floor of ducks should be free from any sharp edges of concrete materials, non abrasive surfaces, free from small stones and plane. Rough flooring also can injure the skin covering, hock joints leads to lameness. With the age of duck the sensitivity of getting injury in the legs increases.

Optimum Temperatures Requirement at Various Stages of Duck Rearing

Temperature is the vital for the survivability of the ducks. Majority of times they usually remain in water. The adult ducks are easily able to maintain their body temperature for presence of feather. At the time of hatching of ducklings temperature plays vital role in duck farming. For this period requirement of temperature is much high to sustain the life due to lack of feather and also due to inability to generate or regulate the body temperature and solely depends upon the external source of heat. This newly hatched duckling is kept in the brooder where ducklings are kept. The ideal temperature of brooder is 33°C at the first day following hatching.

Table: The recommended optimum temperatures for ducks at different ages

Sl. No.	Age in Days	°F	°C
1.	1	86	30
2.	7	81	27
3.	14	73	23
4.	21	66	19
5.	28	59	15
6.	35	55	13
7.	42	55	13
8.	49	55	13
9.	Developing breeders	55	13
10.	Laying breeders	55	13

Source: https://www.vet.cornell.edu/

Grower House

Ducks rearing in intensive and semi intensive system, required space about 1250 cm^2 (i.e. 8 birds/m^2) of floor space per/ grower ducks from 3week to 8 week of age. The space required for waterer is 3.5 cm and feeding is 7 cm. A flock of duckling should not contain more than 500 ducklings together. The floor should be made up of concrete and roof with skillion-roof type. The front covering either is made up of poly weave plastic or hessian blinds during the winter, windy or rainy season.

Run area should be divided to prevent inter flock mixing up of ducks. There should be 60 cm high fences and each duckling should get an area of 2 m^2. Run area better to have water for swimming, which helps them to clean their feather, improve their growth rate. The duckling should be inside the grower house during the period of night.

Wet and dumped litter should be removed from the shed, which generated the much humidity as well as build moulds which are harmful for the ducklings.

Light should be provided to prevent them to get afraid of darkness, easily discover the feeder and waterer, prevent stampeding and cannibalism or feather peaking. The ideal bulb should be of 15 watt and should cover 18 m^2 of floor space.

Ducklings Waterers and Feeders

Water and feeds are provided in the specially made waterers and feeders for the ducklings. Waterers may be of automatic ballcock-operated drinking vessel, or a bell-type hanging drinker. These provide ducklings to immerse their heads in the water and able to drink water and easily acquainted with the water. They should get clean water because they are susceptible for many diseases due to lack of the immunity. Requirement of the space for drinking water is 1.5 cm.

The litter should remain dry in condition. Ducks should be kept in wire mesh or in the concrete floor to prevent spillage of water from the waterer. Due to lacking of water ducklings may show sign of staggering, convulsion. In this condition, it is advised to provide the warm milk rather giving water immediately which may leads intestinal cramps and staggers.

Feed should be provided in the feeders and should be at the level of body so there will be less spillage sand subsequently feed will be free from contamination of droppings. The space requirement for the duckling till 21days is 4 cm.

Fig. Rearing of White Pekin and Khaki-Campble under intensive housing system

4.5 BREEDING MANAGEMENT OF DUCK

Breeder males (drake) and females are selected when they are around 6-8 weeks of age. For maximum fertility drakes should be older by 4-5 weeks than females. In flock mating, a male female ratio of 1:6 to 1:8 is satisfactory with layer types of ducks while ratio should be narrower for meat type of ducks. Eggs for hatching should be collected only from those flocks that are in lay for about 6-8 weeks. Collection of eggs should be started 10 days after introduction of male. Dirty eggs can be washed using warm water at 27°C to which a detergent sanitizer and disinfectant are added which improves hatchability. Hatching eggs should be stored in an atmosphere having a temperature of 14°C to 16°C with a relative humidity of 80%.

4.6 DUCKS EGG

Ducks usually begin laying at about 20-24 weeks of age. The first few eggs will be small and they should not be set for incubation. White pekins start laying eggs at the age of about 26-28 weeks of age and kept economically for egg production up to 40 weeks of age. Optimum egg production and overall performance can be achieved, if breeding ducks are housed in groups less than 250 birds. Litter should be changed on daily basis after most eggs have been laid. Soiled eggs have low fertility. Fertility of soiled eggs can be improved by washed with water containing detergent or sanitizers or disinfectant. Ducks egg shell is little tougher than the chicken eggs, they required more humidity compared to the chicken. Eggs should be fumigated before putting inside the incubator.

4.7 MANAGEMENT OF EGGS

Egg Collection

Ducks are preferred to lay egg in the early morning starting from 4:00am to 7:00 am. Some of ducks goes for laying beyond this time period. Thus it is always good to allow them to leave the house after 10:00am. Although, some ducks are preferred laying in the evening time. It is always better to collect the eggs as early as possible following laying, so they should not get soiled/dirty and chance of minimal damage. Eggs should be collected when ducks left in the run area. The dirty eggs should keep separate from the soiled egg which is covered by faeces as well as the microorganism which has potential to deteriorate the quality of egg.

Egg Cleaning

The collected eggs are covered with litter materials, feather and sometime fully soiled with the faecal material. This egg should be cleaned immediately after collection. It prevents disease spread via faecal material to other healthy and fertile eggs as well prevent spoilage of the egg due to microbial activity. Egg shells are porous through which microorganism easily get chance to take entry. Eggs can be rubbed by to the fine-grade steel wool removes the manure and other materials easily followed by wiped with a clean damp cloth.

In case much dirty egg can be cleaned using warm (about 45°C) water containing sanitizers such as Iodine-based compounds, chlorine solutions and quaternary ammonium compounds. Using cold water for washing leads to contraction of the contents of the egg, subsequently allow dirt and bacteria on the shell to contaminate the egg. Eggs should not be washed with water, it wipe off the wax covering above the eggs and reduce the hatchability. The rest of the eggs should be fumigated in the formaldehyde gas in a fumigation cabinet.

Egg Storage

Eggs are stored in egg cage by putting their broad end above. Egg should be stored in well temperature controlled room. The temperature required is 13-16° C with a relative humidity of 75% for 7 days. Too low temperature leads to death of the embryo. The fertile eggs are stored till the number is enough to full the incubator. With the increasing in storage time the fertility also get decreased, after 7 days fertility drastically reduced. Following storage of 3 week, it is hardly to get live duckling from these eggs. The table purpose eggs can be stored more than 3-4 weeks in low temperature without any problem.

Fig. Storage of eggs

Fumigation

It is a process of surface disinfection which has the capability to destroy harmful bacteria, such as *Salmonella pullorum*. This fumigation is usually done to eggs, incubators and rooms. This disinfection should be performed once in a month.

To disinfect the incubator, 25 g of potassium permanganate and 35 ml of formalin (40%) in earthen pot/ earthenware are used to fumigate 1.0 m^3 of incubator space. The incubator is functionally disabled and kept the pot. Following mixing of the ingredients then the incubator fan should be run for minimum of 10 minutes so that the gas can percolate throughout the incubator. The room should also be fumigated to make maximum sterilization. Person handling the whole process should use all protective gears including respiratory masks. Following fumigation the door of the room should open completely so that the noxious gas can be evacuated naturally. Eggs are generally fumigated during the period of storage and on the days of second candling i.e., day 25 (on day 32 for Muscovies) following incubation.

Egg Incubation

Selection of egg is very crucial to get the optimum hatching from the eggs. Ducks lay egg batch wise and each batch contain 20-30 eggs. The eggs produced in the first batch usually are of smaller in size which is not suitable for the hatching. These are used for table purpose. Also, eggs with underweight, cracked or heavily mottled, or those that have poor shell texture should not be set for incubation.

Majority of farms fumigated the incubator machine with formaldehyde gas (Potassium permanganate with formaldehyde) before setting eggs in to the machine. It should be able to maintain proper temperature and humidity for hatching of eggs. Eggs should be placed putting the pointed end down into the incubator trays. The incubation period for ducks is 28 days while in case of Muscovy require 35 days to hatch. The temperature should be at 37.5°C, relative humidity should be 65-70%. Eggs required sprinkling of warm water at least once daily for initial 15 days and thereafter till 23rd day required 3 times daily. Eggs should be cooled at 32°C for maximum 30 minutes per day from 5th day of incubation onwards. Turning eggs should be done at an angle of 90°C and be monitored within incubator which prevents sticking of the material with the egg shell. Turning should be done every hour when automatic turning facility present in the incubator, while in case of manual turning it should be done 3 times a day.

Monitoring of incubator is very important. High temperature leads to smudgy- burst yolk and low humidity leads to high causes death of the duckling inside the eggs due to unable to hatch out.

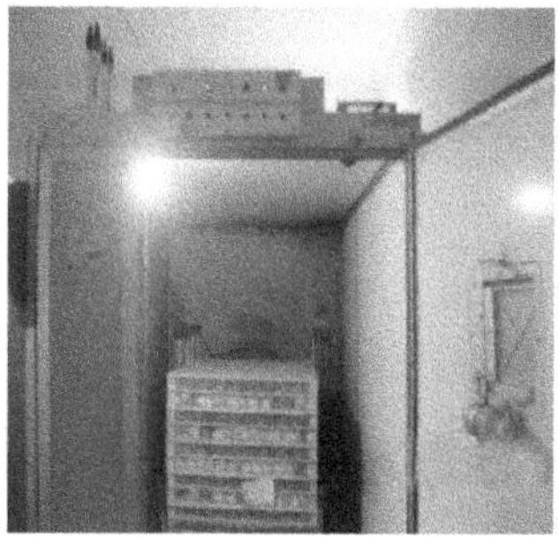

Fig. Setter and Hatcher for artificial incubation of eggs

Candling

Candling is a process to assess for the internal quality of the hatching egg and the embryonic development inside the egg. It is done through using an electric bulb and placing the egg against the light. On 7th day of incubation, eggs are tested for the fertility. In a fertile egg, embryo is visible near the air cell as a dark spot at the large end of the egg and blood vessels radiating from it. A dead embryo generally stuck with the wall of the shell and no radiating masses of blood vessels. The infertile eggs are always clear in appearance. The eggs containing embryo again candled at 24- 25th day of incubation

Fig. Candling of egg

to remove the dead embryo/ ducklings inside the shell and allow all healthy eggs to get proper temperature and humidity for hatching in hatchers. In case of Muscovy ducks candling is done at 32nd day of incubation.

4.8 BROODING

Once hatching is completed, ducklings should be placed in a separate place. Egg shells and dead ducklings should be removed and placed ducklings in the brooder house. In the brooder house, the temperature is controlled at a temperature suitable for ducklings. There should be monitored for overcrowding. The temperature can be maintained by using the electric bulb, heater. The relative humidity should be ranges from 60% to 75%.

On the first week, the temperature of the brooder should be 30°C and gradually has to reduce by 3°C per week till third week of age. Removal of brooder is totally depending upon the climatic conditions of the surroundings area.

Fig. Ducklings Rearing in the Brooder House

Brooder House

Brooder house should be well ventilated, proper arrangements to maintain the constant temperature. The wall height should be 1.2 m from the floor. The floor should be made up of concrete, the depth of litter should be 8 cm. Litter should keep dry by raking the litter and any growth of mouldy litter may lead to death of the ducklings. For 10 days old ducklings required 200 cm^2/bird (i.e. 50 birds/m^2). After two month or on 8th week ducks will be required 1250 cm^2/bird (8birds/m^2). The house should be well protected from the predators. They should allow to run area once they are reaching 2-3 week of age and capable to maintain own body temperature. Monitoring is important for their physical activities during this crucial period of time.

CHAPTER 5

DISEASES OF DUCKS AND PREVENTION

DR. BIPLAB DEBROY AND DR. JOWEL DEBNATH

Domestic breed of ducks are fairly disease resistant as long as they are fed a healthy diet, given plenty of room to exercise and access to fresh water daily. However, there are few infections which can cause significant economic loss, mostly in terms of mortality. Rearing of domestic ducks require man-made or natural water bodies, which upon co-mingling with free-living waterfowl species increases the chance of disease transmission. Diseases of ducks around the world can be conveniently divided on the basis of aetiological agent such as bacteria, virus, fungi etc.

5.1 VIRAL DISEASES

Duck Viral Hepatitis

Duck viral hepatitis (DVH) is a highly contagious and fatal disease of young ducklings which primarily affects bird up to three weeks old caused by duck hepatitis virus (DHV). Three antigenically different viruses namely, duck hepatitis virus (DHV) type 1, 2 and 3, have been found to be associated with the disease. DHV-1 has been reported to infect duck only, is worldwide in distribution and most virulent. Recently, DHV type 1 has been classified as unassigned species in the family *Picornaviridae* whereas, DHV types 2 and 3 are recognised as member of the *Astroviridae* family.

Ducklings are most susceptible to DHV at younger ages and gradually become more resistant to infection as they grow older. The disease is rarely seen in ducklings over four weeks of age. The onset of the disease is very rapid, acute and spreads very quickly through the flock. DHV type 1 once confirmed on site, can remain viable in the environment for weeks to months, which also helps their rapid spread. The key routes of transmission are faecal-oral or by fomite, such as contaminated equipment, vehicles and staff. Vertical transmission does not occur.

DVH type 1 infection generally occurs as acute or per acute infection, with affected birds frequently in good body. Sudden death is the most common clinical sign. Sometimes, sick ducklings develop spasmodic contractions of their legs and die within an hour in a typical "arched-backward" (opisthotonus) position. Mortality occurs up to 90 per cent in young flock and is generally 5% to 10% in an endemic situation. At post mortem, predominant lesions are hepatomegaly with petechial or ecchymotic haemorrhages, secondary bacterial septicaemia (common in older birds) and fatty kidneys.

Differential diagnosis of infections caused by DHV-1, DHV-2 and DHV-3 is difficult by gross and microscopic examination. Recently, one-step multiplex RT-PCR methods are developed which can detect and distinguish between the different types of DHV. Detection of duck hepatitis virus type causing outbreak in an area is important to develop immunisation strategy because, immunity is serotype-specific and does not confer protection against infection with heterologous serotypes.

To prevent the disease, breeder ducks should be vaccinated with a live attenuated duck virus hepatitis vaccine through subcutaneous route, using type 1 virus, provides maternal immunity that effectively prevents high losses in young ducklings. At least three immunisations at every 12 weeks interval during laying period are advisable for adequate passive protection of ducklings. Other preventive measures to take into account such as keeping age groups isolated particularly during the first five weeks of life, avoiding contact with wild waterfowl and rats (being reported as a reservoir of the virus).

Day-old chicks can be vaccinated with modified live DHV-1 vaccine by the subcutaneous route or by foot web stab in a single dose, following which birds rapidly develop an active immunity within three to four days. Hyperimmune serum prepared from the egg yolk of hyperimmunised chickens against DHV type 1, is an effective treatment of affected flock. Hyperimmune serum is applied in the neck by SC route at the onset of the disease.

Definitive diagnosis is by virus isolation by inoculation of blood or organ material into the allantoic sac of an embryonated egg, direct immunofluorescence of the liver, or PCR.

Duck Plague (Duck Virus Enteritis)

Duck virus enteritis (DVE) also commonly known as duck plague, is a contagious acute, sometimes chronic virus infection that naturally affects only ducks, geese and swans of the family *Anatidae*. In DVE infection, significant economic losses incurred due to high mortality and decreased egg production in domestic and wild waterfowl.

The disease is caused by Anatid alphaherpesvirus-1 or DVE virus (DVEV), a member of the *alpha-herpesvirinae* subfamily of the *Herpesviridae* family.

DVE has been reported to occur in all age group of ducks ranging from seven days of age to mature breeders. But mature ducks are most likely to be affected from the disease.

In acute infection, the first signs are often sudden, high and persistent mortality with a significant drop in egg production. In chronically infected flocks, death occurs occasionally. Birds recovered from infection may act as carriers and may shed the virus in the faeces over a period of years.

Severity of infection in a DVE outbreak depends on the species, age and sex of the affected birds, and the virulence of the virus. Clinical signs in affected birds include sudden loss of appetite, watery diarrhoea, nasal discharge, photophobia and ataxia (wings used to support walking or swimming). In ducklings, losses may be lower than in older birds, clinical signs include dehydration, loss of weight and blood-stained vents.

DVE is a vascular disease, because of which, gross lesions are characterised by vascular damage leading to haemorrhage in multiple organs including heart, pancreas, liver, kidneys and lungs. There is also presence of diphtheroid lesions of the mucosal surfaces of the digestive tract and free blood in the body cavity. Haemorrhages can also be found on the mucosal surface of the GI tract, particularly the oesophagus and intestine, which later can progress to necrotic foci in due course of the disease. Necrotic foci may also be found in the liver and cloaca. Microscopic lesions are characterised by vascular damage and subsequent haemorrhage in visceral organs. Epithelial cells of the digestive tract containing eosinophilic intranuclear and intracytoplasmic inclusions are typically present.

DVE in birds more than two weeks of age can be prevented by vaccination with a live attenuated virus vaccine. Fattening or breeding ducks may be vaccinated through subcutaneous or intramuscular route to produce an active immunity. An inactivated vaccine also reported to be as efficacious as modified live vaccine but tested only under laboratory conditions.

Reovirus Infection of Muscovy Duck and Goose

In poultry, avian orthoreoviruses have been reported to cause diseases, such as arthritis-tenosynovitis and stunting runting syndrome. First reovirus infection in Muscovy ducks was described in 1950 in South Africa. However, it is only in 2003, first reovirus infection in ducks was reported.

The disease first appears at 7-10 days of age and may persist in an affected flock until 6 weeks of age. Mortality is higher in young flocks than in older birds. There are no specific signs of this infection. However, acute phase signs include malaise, diarrhoea of sick birds and reluctant to move. Ducks which survive the disease are markedly stunted in growth and generally develop lameness. There is marked swelling of hock, metatarsal joints, including gastrocnemius and digital flexor tendons. Sometimes, swelling is also seen in the synovial bursae. Morbidity and mortality may vary from 10-60 %, and 2-20 %, respectively.

At post mortem, characteristic lesions are observed in liver and spleen: both the organs show elevated cream coloured pin-head size necrotic foci scattered throughout the parenchyma. Both the organs are also enlarged. Pericarditis, arthritis and tenosynovitis are usually seen. In chronic phase of the disease, rupture of the tendon and surrounding tissues in the region of gastrocnemius flexor tendon leading to large area of haemorrhage are observed. Histologically, liver, spleen and myocardium shows foci of coagulative necrosis or granuloma-like foci with necrotic centres surrounded by proliferating macrophages.

Diagnosis of the disease is based on clinical signs and characteristic lesions present in the liver and spleen. Confirmatory diagnosis can be made by isolation of the virus in cell culture and subsequent detection by electron microscopy. Recently, for detection and identification of nucleic acid of avian reoviruses in cell cultures and clinical samples, RT-PCR may be employed.

Presently, no vaccine for prevention of reovirus infection in Muscovy duck and goose has been developed. Although several attempts have been made to develop vaccine but failed to induce protection.

Waterfowl Parvovirus Infection

Goose parvovirus (GPV) and Muscovy duck parvovirus (MDPV) infection causes heavy mortality in goslings and Muscovy ducklings, which may reach 70-100 %. Mortality is highest under one week of age which gradually decreasing with the age. These viruses differ in host range as geese are fully resistant to MDPV infection, but both viruses can cause severe disease in Muscovy ducks. Antigenically, there is a clear distinction between GPV and MDPV with very little or no cross-protection against each other.

The infection in immune-compromised birds may cause significant economic losses up to 6-8 weeks of age. Depending on the age, infection in goose and Muscovy duck may occur either as acute, sub-acute or chronic forms.

Infected birds excrete huge quantity of virus into environment with their faeces during the acute phase of the disease, which acts as source of infection to other birds in the flock. Birds recovered from the disease may continue to shed the virus in the environment and become healthy carriers. Parvovirus being relatively resistant can persist on poorly cleaned and disinfected surfaces for long time, which results in transmission to subsequent flocks. Vertical transmission is also an important route of transmission of the virus.

Parvovirus contains single stranded DNA as nucleic acid. Hence, virus infects mainly rapidly dividing cells; this is why only young birds are affected. Concurrent infection with other immunosuppressive pathogens such as reovirus, cicrovirus and mycoplasmas aggravates the clinical disease.

Clinical signs include prostration and death in acutely affected goslings. There may be reduced feed intake, excessive water intake, swollen eyelids, eye and nasal discharge, profuse white diarrhoea, membrane covering tongue and reddening of skin.

Main post mortem lesions are pale myocardium, swelling and congestion of liver, spleen and pancreas, fibrinous pericarditis, fibrinous perihepatitis and ascites.

Presumptive diagnosis depends on clinical signs, lesions, the appropriate age and species of birds. PCR and serology are the most commonly used laboratory tests for the confirmation of the disease.

Vaccination with live and inactivated vaccines is an effective way to prevent parvovirus infection. Live vaccines containing attenuated goose parvovirus and inactivated vaccines containing the whole parvovirus antigens either in the monovalent or bivalent form can induce rapid immune response and protection. For a stronger and more durable immune response, a booster dose with inactivated vaccine at around 2-3 weeks of age is also recommended.

Hemorrhagic Nephritis Enteritis Virus Infection of Geese (HNEG)

HNEG is a viral disease of geese causing high morbidity and mortality. The first outbreak was reported in 1969 and since then several outbreaks have been reported in major goose-breeding countries. Exact aetiology of the disease was unknown until 2000, which was finally identified and found to be caused by a polyoma virus. Later, based on phylogenetic analysis, the causative agent of HNEG was found to be closely related to but clearly distinct from other polyoma viruses, hence named as 'goose hemorrhagic polyoma virus'.

The outbreaks usually occur between 3-6 weeks of age. However, outbreaks have also been reported in younger (four days old) or older birds (17 to 20 weeks old). The mortality varies between 4 to 67 % and can continue for months. Sometimes two separate mortality peaks may be observed.

Geese affected with HNEG virus may die suddenly without showing any signs. However, geese may sometimes show ataxia, tremors of the head and neck, subcutaneous haemorrhages and the excretion of blood-stained faeces. Geese start dying rapidly once clinical signs develop. Geese if recover from HNEG virus infection are persistently infected.

Detection of polyoma virus nucleic acid by PCR is the method to confirm diagnosis of HNEG. Tissue samples from lung, liver, kidney, spleen, bursa of fabricius and intestinal contents from natural cases of the disease can be used to detect GHPV-specific DNA.

No commercial vaccine against GHPV is available till date mainly due to the difficulties in the propagation of the GHPV in cell cultures.

Circovirus Infection of Geese and Ducks

The diseases caused by genus Circovirus in avian species are relatively little known. Circovirus infection of geese and ducks was first described in 1999 in Germany. They are generally seen during early phase of life in birds. The principal clinical signs include developmental and/or feathering anomalies. There is also significant damage to the lymphoid tissue leading to immunosuppression. The virus impairs both humoral and cellular immune functions which enhances the pathogenicity of co-infecting agents. The outcome of the infection depends on the presence of concurrent infections and other predisposing factors.

The circovirus infected ducks exhibits retarded growth, feathering disorders and increased mortality due to secondary infections with other pathogens such as *Riemerella anatipestifer*, *E. coli* and *Aspergillus sp.* leading to heavy economic loss.

In circovirus infection in waterfowl species, the most important histological changes are observed in lymphoid tissues, commonly in bursa of fabricius (BF). The changes include lympho-follicular hyperplasia, lymphoid necrosis leading lymhocytes depletion and cystic atrophy. The frequent detection of globular or botryoid, intracytoplasmic basophilic inclusions within macrophages considered to be a characteristic feature in circovirus infection.

Circovirus infections are diagnosed on the basis of clinical signs, histopathology features of bursa of fabricius, and demonstration of virus antigen or nucleic acid. PCR is helpful in detecting nucleic acid of circovirus. Molecular epidemiological results suggest that circoviruses are host-specific.

Circoviruses are highly resistant to inactivation, eradication is unlikely to be an option for disease control. Very limited attempts were made to prevent and control diseases caused by circovirus. No vaccine is available till date to prevent the disease.

Avian Influenza

Avian influenza is caused by Type A influenza viruses of the family *Orthomyxoviridae*. Wild ducks are the main reservoir of viruses which can be transmitted to other domestic poultry and mammals, including humans. The influenza A viruses primarily infect free living waterfowls. These viruses are classified by their 16 subtypes of hemagglutinin (HA) and 9 subtypes of neuraminidase (NA) surface glycoproteins. Several duck species are naturally resistant to highly pathogenic (HP) H5N1 influenza viruses. These ducks can shed and spread virus from both the respiratory and intestinal tracts, showing no or very few clinical signs. But in chickens and other gallinaceous poultry, highly pathogenic (HP) influenza viruses cause almost 100% mortality.

In few cases, experimental infection in ducks with HPAI virus showed neurologic signs, including blindness and head shaking with no mortality. Microscopical examination revealed encephalitis in brain, myocarditis with degeneration and necrosis of myocytes in heart and mild lymphoid follicular hyperplasia. When challenged with lethal dose of HPAI virus, ducks showed severe neurologic signs like tremors, incoordination, seizures and even torticollis. However, in 2011, heavy mortality of ducks was reported in a natural outbreak of AI in the north-eastern state of Tripura in India.

Surveillance of the disease and active preparedness for prevention and control of avian influenza is essential. Any unusual mortality in poultry should be reported to the appropriate authority immediately. In the event of avian influenza outbreak, the concerned state and the adjoining states are to be alerted immediately. The bordering states are also need to be alerted in case of an outbreak in a neighbouring country. Several laboratories equipped for faster and accurate diagnosis of the disease need to be established. Veterinary workforce should be trained to handle, control and containment operations. Sensitization of general public or farmers on avian influenza may be done. Import of poultry and poultry products from HPAI positive countries should be banned immediately. Border check posts with neighbouring countries need to be strengthened and should maintain strict vigil on movement of poultry and poultry products.

5.2 BACTERIAL DISEASES

Riemerella Anatipestifer Infection

R. anatipestifer infection is an infectious disease which causes heavy mortality and morbidity particularly in young birds. Ducklings up to eight weeks of age are highly susceptible, and birds infected at less than five weeks old usually die within one to two days if left untreated. *R. anatipestifer* survives in the environment for long time. Route of infection is mainly through the respiratory tract, or through skin wounds. The severity of disease seen depends on the route of entry, where more severe signs are seen with infection through wound than with the oro-nasal route. The incubation period of the bacteria varies between 2-5 days.

Mortality varies between 5% and 75% depending on factors such as age, stocking density and concurrent pathogens. Common clinical signs observed are depression, ocular and nasal discharge, mild coughing and sneezing, diarrhoea, and neurological signs such as ataxia, torticolis and head and neck tremors. The ducks which survives the infection shows stunted growth. Post mortem lesions mainly include cream coloured fibrinous exudate covering serosal surfaces, particularly producing pericarditis, perihepatitis and air sacculitis. The spleen may also be enlarged and mottled. However, both of these lesions are non-specific and require bacterial culture for definitive diagnosis. For culture of this bacterium recommended medium is chocolate or blood agar, with additions of 0.05% yeast extract and 5% newborn calf serum. Swabs for bacterial culture may be obtained from various tissues such as heart blood, air sacs, bone marrow, lungs etc. Beside culture, other diagnostic methods such as immunofluorescence or PCR of affected tissues have been reported to be effective.

R. anatipestifer being bacteria, treatment mainly depends on antibiotic therapy, based on bacterial sensitivity. Vaccination with both live and inactivated bacterin vaccines may be an alternative option to prevent the infection. However, other than vaccination or antibiotic therapy, improvements in basic hygiene with strict cleaning and disinfection protocols are essential in preventing the disease in long term.

Avian Cholera/Fowl Cholera

Avian Cholera is an infectious and contagious disease, caused by a Gram negative bipolar bacteria *Pasteurella multocida*. It causes high mortality in ducks leading to significant economic loss. The ducks of about four weeks of age mainly affected. This disease is associated with poor hygiene and sanitation, and stagnant water in duck pens. The site of infection for *P. multocida* is generally the respiratory tract. The infections may occur as peracute or acute form or as chronic form. In per acute form, sudden death occurs without showing any clinical signs. Few clinical signs such as loss of appetite, high body temperature, thirst, diarrhoea may appear before death in acute cases. In chronic form, suppurative lesions are usually found. At post mortem in acute or peracute form, lesions are dominated by general septicaemic lesions such as petechial and echymotic haemorrhages under the skin, in visceral organs, over the serous surface and intestine. Other common lesions include pericarditis, arthritis, enlarged liver and spleen. The liver becomes copper colored and friable. Pinpoint whitish spots may be seen on the liver. In chronic forms of *P. multocida* infections, suppurative lesions may be found widely distributed, often involving the respiratory tract, the conjunctiva and adjacent tissues of the head. Diagnosis is always dependent upon isolation of the organism. For the detection of subclinical infections, mouse passage of relevant samples is recommended, but polymerase chain reaction and isolation attempts on selective media may represent alternatives.

Antibiotic such as sulfadimethoxine-ormetoprim and chlortetracycline given in feed after sensitivity test are effective. Good sanitation practices are necessary for prevention of this disease. The most effective way to prevent the disease is vaccination. Vaccination of birds, first at the age of 4 weeks and again at 18 weeks, should be done to prevent duck cholera.

Streptococcus infection

Streptococci a Gram-positive non-motile bacterium which form part of the normal intestinal and mucosal flora. The most common species isolated from ducks causing septicaemia are *Streptococcus bovis* or *Streptococcus gallolyticus*. Streptococci, like other commensal such as *E. coli*, are suspected to be an opportunistic pathogen. Streptococcus infection in ducks typically affects young birds, frequently causing sudden death in ducklings usually up to three weeks old. Other clinical signs, especially neurological signs such as tremors, fitting and opisthotonus have also been reported. At post mortem, most common findings are hepatomegaly and splenomegaly with mottling of the spleen.

For definitive diagnosis, culture of the bacteria is must. Streptococcus is usually easily grown on blood agar. Swabs can be taken from the liver, spleen or heart blood. Treatment for streptococcus is mainly based on antibiotic therapy, which should be based on bacterial sensitivity as far as possible. Improved hygiene and managemental practices are necessary to reduce environmental reservoirs of Streptococcus to a minimum.

E. coli infection

E. coli infection is one of the most common bacterial infections of commercial ducks and thus having the largest economic effect. *E. coli* infection can occur in all age group of ducks. Other than a primary causal organism, *E. coli* infection may occur secondary to other infections such as viruses or management problems which may affect the flock.

Signs in affected birds may vary from sudden death, reduced hatchability to birds with their necks pulled into their bodies.

Post mortem signs are similar to those found in the fowl. In acute infections, a congested carcass, lungs and petechial haemorrhages in the heart, air sacs and omphalitis may be observed. However, pericarditis, perihepatitis, enlargement of the liver, air sacculitis and pneumonia is more common in chronic infections. Salpingitis and peritonitis is very common in breeder ducks.

Diagnosis is mainly based on the post mortem findings and isolation of the organism in selective media.

Good sanitation and management practices are important preventive measures. Poor level of cleanliness within the house and drinking water system pre-disposes ducks to this disease. Good sanitation practices are essential at breeding farm or hatchery because contamination of the eggs by *E. coli* will also lead to yolk sac infections. Antibiotic therapy can be effective in combating infections. Commercial inactivated *E. coli* vaccines may be effective in severely affected flock.

Salmonella Infections

Salmonellosis as a primary disease is relatively rare in ducks and is generally caused by serotype *Salmonella typhimurium*. Clinical infections in commercial flocks mainly occur due to poor hygienic and sanitation measures maintained at the breeder farm or hatchery.

Clinical infections usually occur in birds between 3 and 12 days of age. Birds usually show signs of depression and sometimes diarrhoea. During post mortem examination, acute dehydration, enteritis, septicaemia, and the classical caecal cores of white caseous material are observed. Mortality may reach up to 15% in a flock.

Routine serological examination of flocks often reveals quite high levels of salmonella infections with no clinical signs evident. Eradication, although difficult due to re-cycling of the organism from crop to crop, measures include improved rodent control, water sanitisation, improved house hygiene.

5.3 PROTOZOAN DISEASES:

Coccidiosis

Coccidiosis is caused by intra cellular protozoa of the family Eimeriidae. In poultry, most species belong to the genus Eimeria and infect various parts of intestine. The infectious process is rapid (4–7 days) and is characterized by parasite replication and extensive destruction of cells of the intestinal mucosa of the host. Poultry coccidia are generally host-specific and the different species parasitize specific parts of the intestine.

A large number of specific coccidia have been reported in both wild and domestic ducks, of which few are reported to be pathogenic. *Tyzzeria perniciosa* and *Eimeria truncata* are known to be pathogenic in ducks and geese, respectively. Other coccidial infection such as *E. battakhi*, *E. anatis* *Wenyonella anatis* have also been described in ducks in India.

Coccidiosis occurs after ingestion of feed, water, litter, and soil contaminated with sporulated oocysts. Oocyst may also be mechanically transmitted through farm equipment, clothing, insects, farm workers, and other animals. Both clinically infected and recovered birds shed oocysts in their droppings.

Signs of infection vary depending upon the number of sporocysts ingested, the species of coccidia, age of the ducks and previous exposure to infection. Young ducklings (1-8 weeks) may die suddenly without manifestation of any sign. Mortalities may continue for weeks with few ducklings dying each day, if left untreated. Early signs include a tucked-up appearance, blood-stained vent, blood-tinged diarrhoea, and the inability to stand. In older flocks, sub-optimal weight gain may be the only expression of the disease.

The small intestines of ducks infected with *Tyzzeria perniciosa* are often distended and filled with blood and caseous exudate. Mortality in *T. perniciosa* infection may reach up to 30%, and recovered ducks are usually associated with slow weight gain.

Diagnosis is carried out based on clinical signs, necropsy findings and histological study. Fecal samples or intestinal scrapings may be subjected to flotation tests and subsequent microscopical evaluation. To identify individual species, RNA and DNA tests, recombinant DNA techniques or PCR may be used.

To prevent occurrence of coccidiosis, one of the main strategies is to break the cycle of the infection by using the system called "All in – all out". This system allows proper cleaning and disinfection of facilities to reduce microbial load before introducing a new batch of birds and also avoids contact between birds of different ages. Different control strategies have been adopted to avoid heavy economic losses incurred due to morbidity and mortality. Proper biosecurity measures coupled with the use of prophylactics were the first strategies deployed to control this disease. Synthetic drugs such as amprolium, nicarbazin, diclazuril, and toltrazuril were used effectively to control coccidiosis in poultry and other birds for many years. However, the constraint of this strategy in controlling coccidiosis is the emergence of drug resistant isolates of coccidia parasite. Recently, there is a growing trend in the use of natural products such as vitamin E or plants such as oregano to improve the general condition of sick birds suffering due to coccidiosis, probiotics such as diets enriched with lactobacilli to improve the intestinal microbial ecology which also acting as immunomodulators and stimulating humoral immunity used as dietary supplements, whose function varies between the stimulation of immunity. Live vaccine have used especially in commercial broiler farms against *Eimeria spp.* But, vaccination may not completely prevent the disease since there is no cross-immunity between coccidia species.

5.4 FUNGAL DISEASES

Aspergillosis

Aspergillosis also known as Brooder pneumonia is a respiratory disease of ducks caused by a fungal species Aspergillus, most commonly the *Aspergillus fumigatus*. Aspergillosis also occurs in chickens, turkeys, waterfowl, game birds, and other bird species.

Young ducklings, especially during brooding are the most susceptible to infection, though can also develop in stressed or immuno-compromised older birds.

Infection with Aspergillosis occurs through the inhalation of spores, typically from contaminated litter or the hatchery. Infection in young ducklings is usually occurs due to inhalation of spores from a contaminated hatching machine or infected eggs.

Hot and humid conditions favours increase in incidence and severity of the disease. Contaminated poultry bedding is one of the most common sources of infection. High levels of ammonia in the shed also act as aperfect breeding grounds for Aspergillosis.

Aspergillosis mainly affects the respiratory system of infected birds, involving the trachea, air sacs and lungs. Infection occurs as either acute or chronic.

Acute infections are typically seen in young ducklings. Symptoms usually appear in the first 3-5 days after exposure. The most common symptom is rapid, open-mouthed breathing (gasping) due to blockade of air passage. As the disease progresses, young ducklings will gradually show lack of appetite, emaciation, listlessness and dehydration. Swelling of eyes, blindness, and torticollis are also typical of Aspergillosis infections. At post mortem, multiple nodules or plaques in the lungs and air sacs are observed.

Chronic form of Aspergillosis generally affects older or immuno-compromised birds.Ducks with chronic infection may show severe respiratory distress, eye discharge, blindness, and signs of neurological dysfunction.

Aspergillosis can be prevented by avoiding use of mouldy straw and preventing feed from getting wet.

5.5 TOXICOSIS

Aflatoxicosis

Aflatoxicosis is the most prevalent, highly toxic and carcinogenic mycotoxins, having a huge economic impact in poultry rearing. It is a condition caused by aflatoxin most commonly produced by fungi *Aspergillus flavus*, hence, named as aflatoxin. This toxin is also produced by *A. parasiticus* and *A. nominus*. Aflatoxicosis occurs due to consumption of aflatoxin contaminated feedstuffs such as groundnut, maize, rice polish and other tropical feeds on storage. Wet grains, rain and humid weather favour the growth of mould. Ducks are most susceptible poultry species to aflatoxin followed by turkeys, Japanese quail and chickens.

There are 4 types of aflatoxins B_1, B_2, G_1 and G_2, named according to their blue or green fluorescence under UV light. Among these toxins, B_1 is the most toxic and highest in concentration followed by B_2, $G_1$and G_2. Aflatoxin B_1 is the most hepatotoxic, a potent mutagenic and the most prevalent worldwide. Aflatoxin damage immune tissues and suppress innate and adaptive immune responses. So, there is decreased cell mediated immunity and to lesser extent humoral immunity, leading to vaccination failure and lower resistance to secondary infections. Hence, ducks with aflatoxicosis are more susceptible to the bacterial, viral and fungal diseases.

Young ducklings are more susceptible than adults. Adult ducks are relatively resistant to aflatoxin.

Clinical signs include decreased growth rate and poor feed conversion due to reduce protein synthesis, passage of undigested food in the dropping, anaemia of the infected birds due to decrease food intake, decrease egg production in layers, decrease hatchability due to embryonic mortality, lameness, ataxia, convulsions, opisthotonus and death.

At necropsy, liver is swollen, friable, pale or yellow due to precipitation of fat and has haemorrhages. The gallbladder is enlarged and fibrosis of the liver may occur in the late stage. Kidneys and spleen become enlarged. Atrophy of thymus, spleen, bursa of fabricius and testes is evident. Hemorrhages may be present under the skin, on muscles and internal organs.

Ascites and hydropericardium develops in chronic stage primarily due to hepatic damage. Histologically, there is necrosis of liver cells and presence of fatty change. Proliferation of bile duct epithelium is evident.Liver cirrhosis with extensive fibrosis occurs at later stage.

Botulism/Limber neck

Botulism is poisoning caused by the toxin formed in putrefying animal and vegetable matter by the growth of bacteria *Clostridium botulinum*. These bacteria are commonly found in the soil, rotting vegetation, food or carcasses, where they produce a very potent toxin. Chickens, ducks, geese, guinea fowl and many other birds including wild birds are susceptible.

The consumed toxin acts on the nervous system, causing weakness and flaccid paralysis especially in the neck muscles. Botulism more commonly occurs in rainy season, because decomposition of biological material occurs faster, on which birds are more likely to have access.

Birds affected with botulism are found sitting or lying on the ground, hunched over and unable to stand, walk or fly. They are flaccid and weak, and cannot hold their heads up. Commonly, birds are found dead in good body condition, with no evidence of struggling. There is no diarrhoea or nasal discharge or any signs of injury. Several healthy birds in a group often dies at the same time, which reflects the feeding habit of the birds, suggesting congregation and pecking at something they find interesting, with the smallest birds being pushed away. In botulism, birds die rapidly, so most birds do not show any gross lesions at necropsy.

Botulism may be suspected when there is a sudden onset of flaccid paralysis in one or more birds in a group and absence of gross lesions in dead birds. A careful and thorough search for a potential source of the toxin in the area where the birds often roam will provide further prove the fact. Laboratory confirmation of botulism is difficult and not routinely done. However, laboratory examinations may be done to rule out other causes of illness.

There is no specific treatment for botulism. Most birds will die once affected. To prevent the occurrence of the disease, birds should be prevented from having access to possible sources of the toxin such as rotten food scraps or carcasses. Birds should be always provided with fresh food and water. Old or wet food, carcass etc. should be discarded with proper care. If ducks start to show signs of botulism, the remainder of the flock should be shifted to a clean area until confirmed diagnosis is made.

5.6 OTHER DISEASES/CONDITIONS

Sticky Eye/Eye Infection

Eye infections in ducks occur mainly due to scratch, injury during rough mating or debris. Signs of eye infection include closed eyes, bubbling eye, redness or tearing. In such cases of eye infection, gentle cleaning of eyes with saline and making sure the duck has access to a water bowl to submerge her entire head can often clear up the problem. But if the condition doesn't improve in few days, a natural camphor-based solution that can be added to the water or applied to the nostrils may help.

Impacted Crop

Impaction of crop mainly occurs when ducks ingest long pieces of string, plastic, rubber bands or any other indigestible material. If impacted crop is suspected, then the area should be gently massaged and provided with grit, olive oil and plenty of water. Surrounding areas shall be cleaned properly and should be free from any potentially dangerous materials.

Prolapsed Penis/Vent

A prolapse occurs when a portion of the oviduct pushes outside the duck's body while laying egg or penis doesn't retract back to its original position after mating. Most of the time these conditions are corrected itself on its own. But if not corrected on its own, then coconut oil and sugar may be provided for few days to tighten the skin tissue and keep it soft. The area should be cleaned properly. Duck or a drake suffering a prolapse should be prevented from mating and separated while healing in progress. Sometimes, prolapse may be carefully pushed back inside if no improvement is observed for few days. Ample room to exercise and a healthy diet can help prevent prolapses in a flock.

Wet Feather

Wet feather is a condition that occurs when preen gland of duck, which they use to keep their feathers well-oiled and waterproofed, stops working. This condition occurs when ducks do not have regular access to water to swim or kept in a poor and unhygienic condition. As a result, the ducks are not able to stay dry in the rain or water and risking the chance of getting drown. In such cases, ducks are given bath in detergent, then rinsed well and blow dried to remove any dirt and old oil. These ducks should be allowed to go to water bodies only when they have regained their waterproofing. Sometimes, often in severe cases, the duck often has to go through a molt and grow all new feathers to gain waterproofing again.

Wry Neck

Wry neck is a condition where ducks especially ducklings are unable to hold its head up and often not be able to walk properly. It is often fatal in ducklings if not treated. Wry neck can be caused by a vitamin deficiency, blow to the head, or ingestion of toxins. So, supplementation of Vitamin B_1 and E, selenium to their diet can correct the condition.

5.7 VICES

Cannibalism

Cannibalism is a typical sign of salt deficiency. It can be converted to vices due to boredom, overcrowding, lack of ventilation in the house. It can be develop at early age such as 4 week of age.

5.8 PREVENTION AND CONTROL OF VICES

Debeaking, removal of the bean part of the ducks can prevent this cannibalism. To do that deabaking machine can be used. This machine is containing a blade which can be heated by the electricity. This heated blade cut the bill portion and cauterising the cutting edge.

5.9 HEALTH CARE

In comparison to chicken and turkeys, ducks are highly active and less affected by diseases. Due to unhygienic environment and faulty administration or inherent weakness because of breeding, infections occur in ducks. It is important to know about different aspects of healthy ducks prior recognising diseased ducks. Vital characteristics on healthy and unhealthy ducks are enlisted below.

S.I. No.	Characteristics	Healthy Ducks	Unhealthy Ducks
1.	General condition, first impression	Lively	Listless, unusually quiet
2.	Weight	Good	Often light
3.	Growth rate	Normal	Too slow
4.	Eyes	Lively, bright	Listless, dull
5.	Cloacae (genital / anal area)	Large, soft, moist, pink	Shrivelled, dry, discoloured
6.	Skin	Soft, loose	Wrinkled, dry

(Srikanth *et al.*, 2018)

5.10 DISEASE CONTROL AND PREVENTION

Bio Security

It is the foremost important and applicable for all livestock farming. This prevent the induction of organism in the farm premises as well as if reach then bio security measures control the load of the organism at the bio security check points. These check points includes,

1. Restrict the movement of human.
2. Restrict the movement of animals and birds.
3. Restrict to mix up with newly inducted bird with the old flock.
4. Follow proper quarantine for the newly inducted animal/ birds.
5. Regular disinfection of the premises of animal dwelling.
6. Follow up the record of movements of labourers and gather information regarding any outbreak in their villages.
7. Restrict the unnecessary entry of trucks, poultry crates and equipment, they should took entry following appropriate disinfection.

8. Human dealing with the management with the duck should change clothing and boots specially managed for farm premises.
9. Application of disinfectants in the foot baths at the entry points of the farm premises as well as buildings is crucial.

Immunization

Even after the adaptation of preventive measures if organism took entry into the farm premises, the own developed immunity of the animal can prevent them from the mortality. To develop the immunity requires to maintain proper scheduling of vaccination at regular time interval. Vaccines are available for the viral as well as for the bacterial disease for ducks.

VACCINATION SCHEDULE ENLISTED BELOW

Name of the Vaccine	Route	Dose	Age of Duck
Duck Cholera	Subcutaneous	Duckling 1ml	3-4 weeks
		Adult 2ml	After one month of last vaccination
Duck Plague	Subcutaneous	1 ml	8-12 weeks

Environmental Stress

Environmental stress is responsible various biochemical changes within the body. It modulates the immune system and suppresses the immunity of the duck and makes them susceptible for various infections. Even after providing the proper housing system birds used to suffer and make them susceptible for various infections.

Common Medication For Duck

1. Electral: 10g/lt water at the time of arrival of duckling
2. Stresroak: 20g/lt water at the time of arrival and one in every week.
3. Vimeral: 5ml/lt water 5 day continuously every month.
4. Amprolium (Anticoccidiostate): 1g/lt water in normal bird start at 3 weeks of age continued 5-7 days and repeat every 2 month. In case of disease 2-5g/lt water.
5. ISD powder (dewormer): 15g/100birds, start at 3 weeks of age, continued 5-7 days and repeat every 2 month.
6. Tetracycline: 5g/lt water for 5-7 days in case of diarrhea.

CHAPTER 6

SLAUGHTER OF DUCK, GRADING AND PACKAGING OF DUCK MEAT

DR. BIJOY KUMAR SARKAR

6.1 INTRODUCTION

The consumption and demands of duck meat has been in rise gradually in recent years and presently holding the third most widely produced poultry meat across the world proceeding after chicken and turkey. Duck meat holds the second position in poultry meat after chicken in Asia. With the increases demand, the duck meat has become a very acceptable dish in comparison to earlier days as traditional and seasonal dish. In India, duck proportionate around 9% of total poultry population and about 17.8 million, of which 92% are deshi. Ducks are always preferred as laying bird. After 3 to 4 laying, the spent and culled ducks are sold in market for meat purpose. Such ducks faces lack of market demand as a hidden reason of less juiciness, more toughness, less palatability. However the nutritive value with the advances of age (in spent and culled birds) does not show much variation. Considering the market demand and preferences, the constrains can be overcome by further processing of duck meat and its subsequent value addition in forms of meat products which shall enhance the overall acceptability and can boost the socio-economic status of rural / small scale farmers in the villages of India.

6.2 PRE-SLAUGHTER HANDLING

Traditionally when the farm is holding excess livestock, the males are selected first for slaughter followed by older non-productive females. Proper handling of duck before slaughter is very critical. As improper pre slaughter handling may lead to loss by 5-10% during catching and transportation, accounting due to stress and downgrading of the carcasses leading to economical loss to the meat industry. The animal welfare implications are equally indefensible. Improper pre slaughter stress management may alter the physiological, biochemical and physical parameters of duck meat. The stressors exerts deleterious effect on carcass quality includes rough handling of ducks, inversion, partial immobilization, confinement, motion, vibration, noise, air movement, temperature and humidity.

6.3 CATCHING

The most common method of carrying the ducks by catching one leg by hand causes the birds in serious stress and resulting in unacceptable quality such as DFD (Dark Firm Dry) meat. There are four systems for harvesting poultry- loose crates, fixed crates, modules and multiple-floor modules. Loose crates are most often, which are made of wood, wire or plastic used for catching ducks to confine the birds. Empty crates are taken from the lorry into the shed, where a team of catchers fill the individual crates. During summer season to reduce the heat stress fewer birds are filled in each crate compare to other seasons. The birds are caught by one leg and passed into the crate through a flapped opening at the top, which in sufficiently large to insert the bird, but not so large as to allow it to get out easily. After filling the crates birds are carried to lorry for loading in the truck. Loose crate is a flexible system at low capital cost with a high labour requirement. The process has to be careful; otherwise it will result in injury and downgrading of carcass.

6.4 TRANSPORTATION

The critical aspects of transportation by open-sided vehicles are exposure of birds to the existing climate and to high wind speeds (80km/h). Several methods can be adopted to reduce the degree of exposure to adverse weather conditions. In cold and wet conditions, wind increases the chill factor, which provides a beneficial cooling effect. During hot and humidity weather the numbers of birds filled in each crate should be reduced.

6.5 RECEPTION AND UNLOADING

For reception and unloading the area selected must be undercover with sufficient space to contain all the vehicle carrying the birds waiting for unloading. It is better to arrange the vehicles entry to the building at one side and exit at other side after washing to avoid cross-contamination. In warm weather, additional ventilation by fan and evaporative cooling devices are sometimes necessary to regulate the environment to prevent the birds overheating and high humidity. For unloading of loose crates, vehicles with eight stakes at the slaughterhouse must be done one by one and placed on a conveyor system, for carrying them to the hanging station.

After opening the crates the birds removed and hung on the slaughter line. The empty crates are washed properly and kept back to the vehicle. The crates are moved by a combination of driven belt-conveyors and roller track. Nylon chain conveyors are more popular since they do not have to be tensioned and show very little wear.

Some slaughterhouses also practice not to place the bird immediately to the slaughter line when the truck arrives, instead the birds received into a holding area / lariage / fence for some time. Birds are kept in the holding area without feed but adlib clean drinking water is provided. After resting in holding area for 2-4 hours the birds are driven to slaughter line to reduce the stress and to obtain optimum carcass quality.

Appliances and tools required to slaughter duck:

- Sharp knives
- A container of bleach water for sanitizing
- Hatchet or axe
- Container for blood and offal
- Scalding pot and electric or propane burner
- Table
- Container for dressed duck
- Ice or refrigerator for chilling the meat

6.6 PRE-SLAUGHTER INSPECTION

Pre-slaughter inspection is very important for safety of butchers, meat consumers and to control the spreading of diseases. A meat inspector or qualified veterinarian inspects the ducks for presence of any infectious or communicable diseases. Once ducks are found normal and free from any diseases are judged as fit for slaughter and should be sent for slaughter.

6.7 SLAUGHTER PROCESS

Some of the conventional household methods for duck slaughtering are as follows:

a) Bleed out the duck by nicking arteries on each side of the neck,
b) Decapitate the duck quickly with a hatchet,
c) Brain the duck with a sharp knife,
d) Break the duck's neck,
e) Shoot the duck in the head.

For large scale slaughter, the birds may be hung by legs at a convenient height for bleeding or they may be placed in killing cones, fastened to a wall or in a rack. Openings at the bottom of the cones allow the fowl's heads to protrude. Commercially, most birds are hung on shackles attached to an overhead conveyer.

6.8 STUNNING (IMMOBILIZATION)

For duck, Immobilization is not critical for manual slaughtering, but necessary for slaughtering by using automated equipment. More recently, stunning has been used primarily from animal welfare perspective as a means to minimize the pain and suffering associated with the slaughter process. Immobilization must be done before slaughter of duck in order to reduce thrashing about before and following the bleeding operation, reduction of possible bruising and discoloration. Various methods of immobilization are such as gas-stunning or electrical stunning is employed. Electrical stunning is usually carried out in an electrically charged water bath by dragging the heads of the birds through water in which an electrode is submerged. The shackles of the killing line simultaneously touch an earth electrode, causing an electrode current to run through the whole body of the bird. In case of gas stunning the birds can be left in the cages where they are stunned by CO_2, argon or their mixture and then removed from the crates and placed on the shackle line. In this case, unloading should be done immediately, so that no time is allowed for the ducks to regain consciousness.

6.9 STICKING AND BLEEDING

Ducks are most commonly slaughtered by severing left jugular vein and carotid artery. Ducks may be immobilized or controlled by placing them in killing cones of proper size to prevent bruised, broken and discoloured wings. Use of cones also saves the operator from painful beatings from powerful thrashing wings. Best practice of bleeding operation must be followed with most feasible bleeding time prior to scalding in order to avoid the incomplete bleeding as to reduce discoloration of carcass. The bleeding time of duck is generally 2 to 5 minutes depending upon the body size, stunning method applied and time interval between stunning and bleeding.

6.10 SCALDING

Scalding is the process of immersing the birds in warm water to loosen the feather. After successful bleeding the birds are scalded by immersed in hot water or by spraying hot water. Scalding tanks are mostly used than spraying-scalding systems. During scalding the ducks are immersed in hot water for up to 3 minutes in the scalding tank (60-65°C). The water must be circulated continuously by means of pumps or agitators at the centre of the tank and overflow. In small plant, scalding can be done manually by placing the carcasses in and removing them from a scalding tank, whereas in large plants, it is done in a continuous manner employing a single stage or multistage scalding bath while the ducks are suspended from a moving shackle line. Adequate agitation of water and uniform water temperature are essential to ensure good feather removal. To maintain the hygiene of the carcass counter-flow design is preferred where clean water being introduced at the exit end of the tank and water flow toward the entrance where the more contaminated ducks are introduced. Multistage scalding tank operation can further reduce contamination problem, which can include 2-4 water baths and the carcasses are moved from the initial, more contaminated bath to the cleanest bath at the end.

6.11 DE-FEATHERING, WAXING AND SINGEING

Ducks are much harder to pluck than chickens. De-feathering process includes picking, pinning and singeing. De-feathering may be done by hand with the birds hanging at a convenient height, or mechanically by using de-feathering machines. In large processing plants, feather removal is done by mechanical pickers/ pluckers/ de-feathering machine equipped with rubber fingers that rub the feathers off the carcass. In a continuous operation, it is done while the carcass is hanging upside down and moving forward in between 2-3 sets of drums covered with rubber fingers. Elasticity and length of fingers varies depending upon the task required and speed of the machine. In small scale operation, de-feathering is done either in batch type equipment where the carcasses are placed in a rotating drum equipped with rubber fingers or by handpicking of the feathers. The operation of pinning consists of hand removal of any remaining pin feathers by use of short-bladed pinning knives. Pin feathers of duck carcass can be removed by wax dipping after mechanical or hand picking. The rough-picked birds are coated with a melted wax. Duck carcasses are suspended in hot wax, followed by cold water immersion to harden the wax, which is later peeled off in large pieces to pull off the pinfeather. In singeing, the hair-like feathers remaining on the birds after picking are removed by passing through gas flames continuously or burning intermittently by means of automatic controls. The carcasses are then rinsed to remove all the soil left after de-feathering and singeing process. In large processing plants, a set of high pressure spray nozzles are used to remove all the debris from skin while carcasses are moving on the shackle line.

6.12 OIL GLAD AND FEET REMOVAL

Oil gland and feet can be removed manually or by automatic equipment. A set of metal bars or rails are used to remove the oil gland. Along the shackle line duck carcasses are positioned at a certain angle to allow the rotating blade of metal bars or rails to cut off the oil gland from tail area. Removal of feet can be done manually, but in modern plant automated equipments are used. Knee joint is positioned at an angle by guiding bars along the shackle line and cut off while passing by a circular rotating blade. It is very important to cut in between the bones and not through the bones to avoid the appearance of dark/red colour in the chilled carcass and dark/ black colour after cooking. After removing feet, the carcasses are re-suspended to another line from the knee joint, which also assists in reducing contamination as dirty shackle line is replaced with cleaner one.

6.13 EVISCERATING

Before eviscerating the carcass must be washed and cleaned. Evisceration may be performed manually by a knife and a pair of scissors or semi-automatically or fully automatically by using a circular cutting blade and a scoop like arm to remove the viscera. Special care must be taken not to pierce the viscera and contaminate the carcass.

Manually once the carcass skinned or plucked, cut around the vent with knife, but care must be taken not to puncture the intestines. Start on the belly side, between the rib cage and the vent and make a shallow incision into the flesh. Use fingers to pull this open, pulling the meat out and away from the intestines. The gizzard will be attached to the other end of the intestines. Stuff hand into the body cavity to pull it out. Most likely pull out the gizzard, liver, and gall bladder in one big mess. Next, find the heart and possibly the crop and trachea. The crop usually pulls away from the gizzard and liver and may be removed from the neck area. The lungs are embedded in the rib cage and are a bit harder to pull out. Sweep fingers between the ribs and pull the tissue out.

In automatic operation, a mechanical vent-cutter is sometimes used. This has a central pin, which is put into the vent. The vent is then sucked by vacuum and cut by a revolving, cylindrical blade. The connection with the intestine is not severed. The initial cut is enlarged with scissors to allow the viscera removing.

Intestines must be washed out the contents and cleaned for consumption. For skinning of bird, slip the knife at neck region under the skin where head to be removed. Lift and slice the skin along the belly.

6.14 POST-SLAUGHTER INSPECTION

Usually a qualified government veterinary officer carried out the inspection of individual carcasses with attached and detached viscera to reveal disease/ problem as to ensure that only disease free wholesome birds will get to the market. The inspection area should have adequate bright light, hand washing facility, a rake for placing suspected birds and a bin for condemned birds. In automatic operation, line speed must be adjusted as to inspect every duck by the inspector or more than one inspector can be engaged for same line. Sometime, a mirror is installed on the back side, to facilitate the inspector to view both sides of carcass without touching carcasses.

6.15 GIBLET HARVESTING

The edible viscera includes heart, gizzard and liver are separated and washed. Giblets (Gizzard, liver and heart) are salvaged and washed manually or automatically in a separate line. The gizzard will need to be cut open, cleaned and yellow kaolin lining must be peeled off before cooking. Mechanical equipment used for peeling consists of two rollers with opposite direction. After peeling gizzards are inspected, washed and immediately chilled. The hearts and livers are collected, inspected, washed and chilled. Sometimes, neck is removed from carcass and included with giblets. The chilled giblets are than packed in a paper bag for further processing or cold storage or to sell in bulk.

6.16 LUNGS, HEAD AND CROP REMOVAL

Although lungs, head and crop are generally removed after inspection, sometimes one or all of these organs removed prior to inspection. Lungs can be removed manually by using a rake-like device or by using a vacuum gun in semi-automated process. The lungs and any other materials remaining within the carcass are removed with a special hand-tool or lung gun by suction. Removal of head can be done manually by using a knife or a pair of shears or automatically by using head puller. Removal of head by using head puller also pulled out oesophagus, trachea and crop together.

6.17 WASHING

Although, washing of ducks are carried out at different points along with the processing line, one of the most common points is prior to chilling. In automatic operation, a device with multiple spray points is used to cover the outside washing of carcass. A retracting shaft equipped with high pressure nozzle is used to wash inside. High pressure low volume nozzles with proper positioning are effective in removing debris. Washing of carcass should start with upper part and gradually the lower parts to minimize the contamination of carcasses. Permitted bactericidal, such as chlorine and organic acids can be used. Chlorine in level up to 20 ppm is most commonly used chemical in washing of carcasses.

6.18 CHILLING AND PACKAGING

After washing the carcass thoroughly, chill the carcass as quickly as possible to prevent contamination of carcass. The cleaned carcass can be kept in the refrigerator for 24 hours, so the muscle tissue will be tender. Soaking in a brine solution also can be followed for most tender meat.

The most common methods of chilling duck carcass include water-immersion chilling, air chilling and spray chilling. In immersion chilling, long chillers with a counter-flow of cold water supplemented with crushed ice are used to bring the temperature of carcasses to about 4-5°C within 30-75 minutes. The counter-flow technique provides efficient way of cooling the carcasses and also assists in improving hygienic conditions. Air chilling of duck carcasses is achieved by using cold air as chilling medium, maintaining high air humidity and wetting the carcasses at some points along the chilling process. Air chilling provides drier final product with minimal drip loss and better microbial property of the product. In spray chilling cold water is continuously sprayed over the carcasses while moving on the shackle line. Moisture pick up in spray chilling is less than water immersion chilling but more than the air chilling. Method of chilling duck carcasses are depends on the market demand, water availability and cost, electricity cost and availability of capital investment.

6.19 WEIGHING, GRADING AND PACKING

After chilling of duck carcasses are usually weighted, graded and packed or further processed. Automated weighing machine and sophisticated computer systems are now used to weigh the carcasses in large processing plants. Grading is not usually mandatory but it done in most of the large markets to facilitate sales. Grading serves to segregate products into standardized groups with common characteristics, like appearance, physical properties or edible portion. Grading of duck carcasses can be done as follows:

Grade	Grade A	Grade B	Grade C
Conformation	Normal	Normal	Abnormal
Keel bone	Normal	Normal	Slightly crooked
Back	No deformities	No deformities	Carcass is whole
Leg & wing	No broken bone allowed	No broken bone allowed	Broken bone allowed
Fleshing	Plump, Full breast On both sides of Keel bone	Moderately plump	Slightly plump
Fat (abdominal and subcutaneous fat)	Properly deposit	Slightly much deposit	Too much or too less deposit
Pins	No pins left	Some pins are left	Many pins are left, (scattering)

[Table Contd.

Contd. Table]

Grade	Grade A	Grade B	Grade C
Skin and bone are damaged	No	No	Bruised skin or broken are allowed
Color	Yellow or white, No prominent discoloration	Yellow or white, No prominent discoloration	Bloody, spotted and discoloration
Maturity	Suitable	Suitable	Too old or too young
Carcass weigh	2.5kg	2.5kg	exceed 2.5kg

For freezing the dressed carcass must be wrapped or packed and freeze immediately. Remove the bird from the freezer two or three days before cooking and allow it to thaw completely in the refrigerator before cooking.

DUCK SLAUGHTER PROCESS

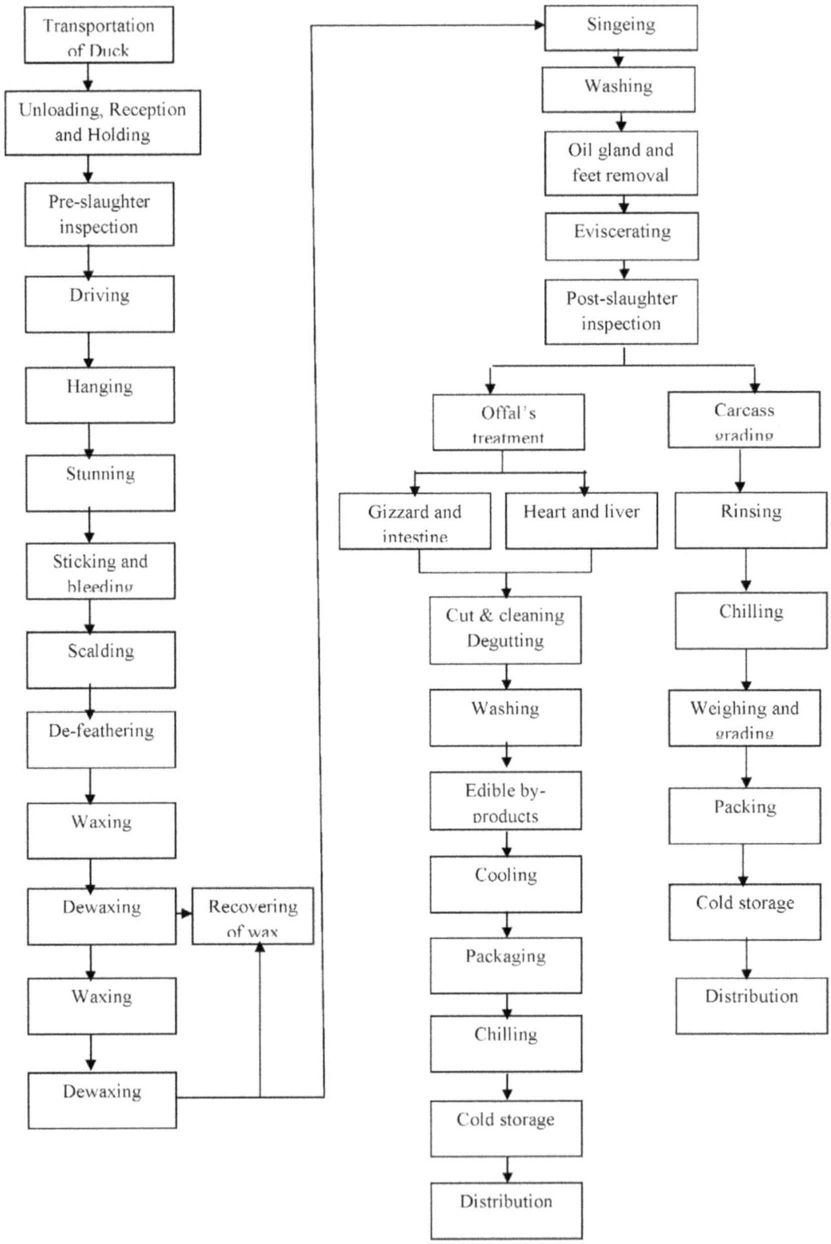

CHAPTER 7

NUTRITIONAL IMPORTANCE OF DUCK MEAT AND EGG

DR. JOWEL DEBNATH

7.1 DUCK MEAT

Duck is a waterfowl and has a diverse physiology to that of other poultry. Demand of duck is very high in many area of the world, especially in Asia. Duck meat production is mainly based on different crossbreds commercial of Pekin. Slaughtering procedure for duck is quite similar to chicken. In comparison to chicken, duck has high red muscle fibre in breast (Smith *et al.*, 1993) and hence considered as red meat. The proximate composition (%) of chicken and duck breast meat enlisted below.

THE PROXIMATE COMPOSITION (%) OF CHICKEN AND DUCK BREAST MEAT

Source of Meat	Proximate composition (%)			
	Moisture	Protein	Fat	Ash
Chicken breast	75.47	22.04	1.05	1.07
Duck breast	76.41	20.06	1.84	0.92

Duck meat is highly flavourful and contains more nutrients. Duck meat is a high source of protein and healthy fat, micronutrients such as selenium, iron, and niacin. Information on nutrition for one 3-ounce (85g) roasted skinless Pekin duck breast, is provided by the U.S Department of Agriculture (USDA) is given below.

DIFFERENT COMPONENTS OF DUCK MEAT

Parameters	Value
Calories	119KCal
Protein	23.5g
Fat	2 g
Sodium	89 mg
Carbohydrates, Fiber ,Sugars	Negligible

7.2 DUCK EGG

Duck egg also is an excellent source of animal protein. Duck egg contains high amount of protein as compared to chicken egg. Different comparative proximity of Duck and Hen egg albumens components are presented below.

COMPARATIVE PROXIMITY OF DUCK AND HEN EGG ALBUMENS COMPONENTS

Parameters(%)	Duck Albumen	Chicken Albumen
Moisture	86.19	88.48
Protein	12.15	10.03
Crude Fat	0.03	0.01
Carbohydrate	0.91	0.75
Ash	0.72	0.73
Fiber	Nil	Nil

CHAPTER 8

INTEGRATED FARMING

DR. JOWEL DEBNATH

8.1 INTEGRATED FARMING SYSTEM (IFS)

Integrated Farming System (IFS) is a amalgamation of numerous systems, it efforts to boost up farmers income using natural resources on sustainability basis which can be obtained by integrating crop husbandry with associated enterprises is called as IFS. Duck is suitable for integrated farming because various ways it allows mutual benefit between different farming systems. Duck farming could be combined with crops, livestock, and aquaculture for improved yielding by taking benefit of farmer's resources without hampering the environment and gain highest profit.

8.2 OBJECTIVES OF INTEGRATED FARMING SYSTEM

- Proficient recycling of farm and animal wastes
- Reducing the losses of nutrient
- Increasing the efficient use of nutrient
- Implementation of well-organized cropping systems and crop rotations
- Balancing the combination of farm enterprises.

8.3 DUCK REARING INTEGRATED WITH FISH POND

In duck cum fish farming, surplus and excreta from the duck shelter can be used after reprocessed for fish culture. This type of natural processing mechanically gearing up the production of natural food in the ponds and subsequently develops the fish production. More revenue can be generated by duck cum fish farming and also beneficial for the farmers. Accumulated waste can be used as good manure and also distribution of waste can be equal, when ducks are allowed to access freely in the fish pond. As a result of this procedure, there is a reduction of expenses for manure, feedstuff, and complementary nourishment for fish. Availability of ducks in the field inhibits the growth of aquatic weeds and promotes ecological production of the ponds. Oxygen quantity in the ponds is enhanced by swimming action of the duck. As ducks consume the weeds, insects, larvae, worms etc. accumulate in the pond; hence there is no requirement to add extra feed to them.

In this farming system, length of the fish plays vital role for overall improvement of the pond and successful farming. Consequently, fishes with 10 cm length merely to be reserved since fishes less than this length may be eaten by the ducks. Fish seeds can be reserved at the rate of 10000 numbers/ha for overall improvement of the pond. Stocking density may vary as per nature of the fish pond and availability of fish seeds. In integrated farming system, growing of the ducks depends on type of the species and egg laying capacity. Proper care and management is necessary to get more meat and egg from duck cum fish farming. Generally for one hectare water area, 300 ducks could be reared economically.

8.4 FISH SELECTION FOR INTEGRATED DUCK-FISH FARMING

In integrated livestock-fish farming phytoplankton, zooplankton and bacteria consuming fishes are considered as most suitable. Common diseases and parasites resistant fishes should be selected.

Criteria for Fish Species Selection:

- High organic and low oxygen levels tolerating fish species should be considered.
- Common diseases and parasites resistant fishes should be selected.
- The species combination ratio should be managed according to the available amount of feed stuff and manure.

FISH SPECIES WITH STOCKING RATIO (%) AS PER TROPHIC NICHE

Trophic niche	Fish species	Stocking ratio (%)
Surface feeder	Silver Carp	20
	Catla	20
Mid-water feeders	Rahu	20
Bottom feeders	Grass carp	10
	Mrigal	15
	Common Carp	15
Total		**100**

(Biswas, 2015)

8.5 PREPARATION OF POND

Small farmer can easily manage a pond of size 0.5-1.5 bigha. Depth of water of the pond in dry season ponds should retain 1.5 to 2 m. Below 1.5 m depth of water could be caused of organic overloading in summer and it could be danger for fish. Hence, considerable depth of water is important.

8.6 SUITABLE DUCK BREEDS FOR INTEGRATED DUCK-FISH FARMING

Khaki Campbell, Indian Runner or cross with local indigenous ducks could be chosen for fish cum duck farming.

8.7 DUCK HOUSE

A duck night shelter along with surface and a small bridge should be made over the pond for feeding of duck and collecting eggs. Average space requirement is 0.3-0.5 m^2/bird in duck shelter house. Cleaning of the dry and wet runs should be done once in a day from shed over the pond and allow entering the waste in to the pond. Appropriate ventilation is recommended for successful duck cum fish farming.

8.8 DUCK AND FISH FEED

Poultry feed and fine rice bran is usually used in addition as duck feed @ 100 - 120 gm feed/day/duck as natural food available in the pond is not enough for proper growth of duck. Tadpoles, juvenile of frogs, dragon fly larvae and various other organic materials are available in the pond which are consumed by ducks and protein content in natural food organisms of the ponds is very high. Therefore, duck reared in fishponds gives more eggs in comparison to duck reared in exclusively intensive housing system.

Duck excreta helps in fertilizing the pond water to produce fish food organism - phytoplankton and zooplankton and on average 60 - 65% of food requirements of farm fishes are fulfilled by feed offered to the ducks and duck dropping. Therefore, extra feeding to the cultured fish is not required in integrated duck-fish farming.

8.9 ADVANTAGES OF DUCK-CUM-FISH FARMING

- Duck's droppings directly utilizes by some fishes as food.
- Ducks keep water plants in check.
- Duck and fishes combined utilize the pond water as requirement.
- Ducks escaping the nutrients from the pond mud at the bottom through dabbling and manage water plants and increases the pond productivity.
- Duck meat and eggs and fish can be produced from a single cultivation process.
- From the pond aquatic weeds, insects, molluscs, etc gives 50- 75% of feed requirement of the ducks.
- Time, labour and cost saving for the farmers.

8.10 INTEGRATED RICE-DUCK FARMING

Duck helps in growing of rice plant by managing the insects and pests and utilize the paddy field as food source.

Reduces Use of Pesticides

Duck change the ecological niches and form new system with new food chains and food webs and improve the living environment as favourable for rice growth and enhance its resistance. A significant decrease in numbers of pests is found in rice-duck cultivation system which reduces the adverse effect of pest on rice cultivation. Nitrogen fertilizer application reduces the pests and spider but duck available in paddy field can manage it very easily.

Reduces Fungus and Bacterial Growth in Rice–Duck Cultivation

Activity of duck on paddy fields can change the living environment for micro organisms which acts adversely on rice growth. Mercaptan, phenolic acids and organic matter are extracted from duck droppings which have antagonistic effect on bacterium and inhibit sheath blight of rice. In general, to control plant pathogenic fungus *Rhizoctonia solani* and other microorganisms inoculate with the mud-capped rice sheath, there is a need of chemical treatment but duck can eliminate the issues. There is almost no difference between chemical control and rice duck co-culture effect in sheath blight or rice blast management. Acting as indirect agent, ducks prevent several rice microbial diseases.

Reduces Use of Chemical Fertilizer

Heavily using of fertilizers could have very bad impact on rice to utilize as food. So for safety in food purpose, rice–duck complex ecosystem can be a positive and effective system to bypass these problems. Improvement of soil nutrients can reduce the use of fertilizers. Duck promote aeration conditions to absorb nutrients by rice plant as well as reduces algae and weeds by lower the consumption of fertilizers. Duck dropping contain 7.1 $gkg^{-1}N$, 3.6 $gkg^{-1}P$ and 5.5 $gkg^{-1}K$ on average and micro elements like Fe, Mn, B and Ca etc. Micro elements present in duck droppings accelerate the decomposition of nutrient towards improvement of soli nutrients.

Reduces Greenhouse Effect

Rice field are responsible for releasing CO_2, CH_4 and N_2O largely have significant impact on global greenhouse. Emitting a noticeable amount of methane (5–19%) globally and N_2O

(7–11%) nationally is a current issue to reduce the release of green house gasses from paddy field. Study shows that a duck activity in the paddy field reduces greenhouse effect by reducing the release of methane as 60% of the paddy fields greenhouse gases are CH_4. Duck increases the soil aeration so methane oxidation increases thereby lower the production of CH_4. Duck activity on paddy field lower the action of methanobacteria, methanogens and lower down the emission of methane.

REFERENCES

1. Ash, W.J.(1976). "Raising Ducks", U.S.D.A., Farmer Bulletin No. 2215.
2. Asok Biswas. (2015). Fish Duck Integrated Farming. https://www.researchgate.net/publication/287595864
3. Banerjee, S. (2013). Morphological Traits of Duck and Geese Breeds of West Bengal, India. Animal Genetic Resources, 52: 1-16.
4. Basic Animal Husbandry Statistics (BAHS). (2019). Govt. of India
5. Byron Stein. Introduction to Commercial Duck Farming. Retrive on 31st Dec. (2020). https://www.dpi.nsw.gov.au/animals-and-livestock/poultry-and-birds/species/duck-raising/commercial-duck-farming
6. Das, S., Rahman, S., Das, K. S., Kalita, G and Tolenkhomba, T.C. (2020). A Socio-Economic Analysis of Duck Farmers of Tripura. International Journal of Livestock Research, 10(10): 144-151.
7. Dean, W.F. (1986). Nutrition of the Pekin Duck in North America: An Update. Proc. 1986 Cornell Nutrition Conference, Ithaca, N.Y., pp.44-51.
8. Dougherty E. (1956). Yearbook of Agriculture: 496-499. Available from: https://naldc.nal.usda.gov/download/IND43894744/PDF.
9. Duck Management and Disease Prevention. Krishi Vigyan Kendra. Lohit, Namsai Dist., Momong Arunachal Pradesh Pin-792102

10. Elkin, R.G. (1987). A Review of Duck Nutrition Research. World's Poultry Science 43, No. 2:84-106.
11. Farrell, D.J., and P. Stapleton, eds. (1986). Duck Production Science and World Practice. Printed by the University of New England, Armidale.
12. G. D. Butcher, J. P. Jacob, and F. B. Mather. Common Poultry Diseases. Accessed on 31st Dec. (2020). https://edis.ifas.ufl.edu. https://edis.ifas.ufl.edu/pdffiles/PS/PS04400.pdf
13. Gajendran, K and Karthickeyan, S. M .K (2009). Status of indigenous ducks in Tamil Nadu, India: a traditional perspective. *Livestock Research for Rural Development. Volume 21, Article #175.* Retrieved December 8, 2020,
14. Gracey, J. F., Collins, D. S and Huey, R. J (1999). Meat Hygiene (10th edn), W. B Saunders Company Ltd.
15. Gerhold Jr. R.W. (2014). Overview of Coccidiosis in Poultry. Available from: https://www.msdvetmanual.com/poultry/coccidiosis/overview-of-coccidiosis-in-poultry
16. http://ecoursesonline.iasri.res.in/mod/page/view.php?id=129722
17. http://www.cpdosrbng.kar.nic.in
18. http://www.kiran.nic.in/poultry.html#:~:text=The%20indigenous%20chicken%20breeds%20include, Runner%20 and %20Brown%20water%20ducks.
19. http://www.lrrd.org/lrrd21/10/gaje21175.htm
20. https://agritech.tnau.ac.in/animal_husbandry/animhus_duck_breed.html
21. https://www.dpi.nsw.gov.au/animals-and-livestock/poultry-and-birds/species/duck-raising/housing-design
22. https://www.vet.cornell.edu/animal-health-diagnostic-center/programs/duck-research-lab/about

23. https://www.vet.cornell.edu/animal-health-diagnostic-center/programs/duck-research-lab/publications
24. Islam, R., Mahanta, J. D., Barua, N., & Zaman, G. (2002). Duck farming in north-eastern India (Assam). World's Poultry Science Journal, 58(4), 567-572.
25. J. F. Huang and C. C. Lin. (2011). Production, Composition, and Quality of Duck eggs. Wood head Publishing Limited
26. Jean E. Sander. Fowl Cholera. Accessed on 31st December. (2020). https://www.merckvetmanual.com/poultry/fowl-cholera/fowl
27. Johnston, A. (2009). Current Diseases of Ducks and Their Control. International Poultry Production. 5(6):11-13. Available from: http://www.positiveaction.info/pdfs/articles/pp15.6p11.pdf
28. Kamal, R., Dey, A., Chandran, P. C., Mohanta, R. K., Giri, S. C., Mohanty, S. and Barari, S. K. (2019). Phenotypic and Morphometric Characterization of Desi Duck of Odisha.
29. Kavitha K, Manohar Raj G, Vairamuthu S, Ramamurthy N. (2017). Comparative study of egg quality traits in white pekin and indigenous ducks of Tamil Nadu. International Journal of Science, Environment and Technology. 6(6):3520–3523
30. King E. (2017). Infectious Diseases Affecting UK Commercial Duck Species. Vet Times. March 6; 1- 11. Available from: https://www.vettimes.co.uk
31. Kromm, M and Lighty, M. (2020). Aspergillosis in Poultry. Available from: https://www.msdvetmanual.com/poultry/aspergillosis/aspergillosis-in-poultry
32. Kunnath, S. K. and Kumar D. A. P. (2018). Duck Farming-An Alternative to Poverty Alleviation. Indian Farmer. 5(03):258- 268

33. Md. Shawkat Ali, Geun-Ho Kang1, Han-Sul Yang, Jin-Yeon Jeong, Young-Hwa Hwang Gu-Boo Park and Seon-Tea Joo. (2007). A Comparison of Meat Characteristics between Duck and Chicken Breast. Asian-Australian Journal of Animal Science. 20(6) : 1002 – 1006

34. Mead G C (2004). Poultry Meat Processing and Quality, Wood head Publishing Limited.

35. Ming-Tsao Chen, Aimin Jiang and Herbert W. Ockerman (2012). Duck and Goose Meat Product Processing Technology. Fulin Publisher.

36. Morduzzaman, M., Bhuiyan, A. K. F. H., Rana, M., Islam, M. R., & Bhuiyan, M. S. A. (2015). Phenotypic characterization and production potentials of Nageswari duck in Bangladesh. Bangladesh Journal of Animal Science. 44(2), 92-99.

37. Mountney, G. J. (1989). Poultry Products Technology. Food Products Press, New York, NY.

38. Mullin, J. (1987). Game Bird Propagation – Wildlife Harvest System. Arrowhead Hunting & Conservation Club, Goose Lake, Iowa 52750.

39. National Research Council. (1994). Nutrient requirements of poultry. National Academy Press, Washington, D.C.

40. OIE Terrestrial Manual. (2018). Duck Virus Enteritis. Available from: https://www.oie.int/fileadmin/Home/eng/Health_standards/tahm/2.03.07_DVE.pdf

41. Pan Long, Huang Huang, Xiaolan Liao, Zhiqiang Fu, Huabin Zheng,, Aiwu Chene and Can Chen. (2013). Mechanism and Capacities of Reducing Ecological Cost Through Rice–Duck Cultivation. Wiley Online Library

42. Parkes H and Shilton C. (2019). Botulism in chickens, ducks and other poultry. Available from: https://nt.gov.au/__data/assets/pdf_file/0010/233569/botulism-in-chickens-ducks-and-others.pdf
43. Pattison, M., McMullin, P., Bradbury, J and Alexander, D. (2007). Poultry Diseases. 6th edn. Saunders Ltd.
44. Phookan, A., Das, B., Das, A., Islam, R., Sharma, M., Bharali, K., & Basumatary, K. (2018). Morphology, Morphometry and Certain Egg Quality Traits of Indigenous Ducks of North Eastern Region of India. International Journal of Chemical Studies. 6(2), 3131-3133.
45. Pundir, R. K., Niranjan, S. K., Behl. R. (2013).Sustainable Utilization of Indigenous Animal Genetic Resources of India. National Bureau of Animal Genetic Resources, Karnal- 132.
46. Rajesh Singh and Ritesh Pandey. (2021) . Modern Technique & Method of Fish Cum Duck Integrated Farming in India. https://www.pashudhanpraharee.com/author/pashudhan praharee
47. Reddy D. V. Applied Nutrition (Livestock, Poultry, Rabbit and Laboratory Animals). Third Edition. Oxford and IBH Publishing.
48. S.K. Ranjan. Animal Nutrition in Tropics: Fourth Revised edition. Vikas Publishing House Pvt. Ltd.
49. Sai Barbut (2001). Poultry Products Processing- An Industry Guide. CRG press.
50. Sams A (1999). Commercial implementation of post mortem electric stimulation. Poultry Science. 78:290.
51. The poultry site. (2011). Major viral diseases of waterfowl and their control. Available from: https://www.thepoultry site.com/articles/major-viral-diseases-of-waterfowl-and-their-control

52. Tirath S. Sandhu. Duck Health Care. Accessed on 29th December. (2020). https://www.vet.cornell.edu/animal-health-diagnostic-center/programs/duck-research-lab/health-care

53. Veeramani, P., Prabakaran, R., Selvan, S. T., Sivaselvam, S. N., & Sivakumar, T. (2014). Morphology and Morphometry of Indigenous Ducks of Tamil Nadu. Global Journal of Medical Research, 14, 17-20.

For Product Safety Concerns and Information please contact our EU representative GPSR@taylorandfrancis.com
Taylor & Francis Verlag GmbH, Kaufingerstraße 24, 80331 München, Germany

www.ingramcontent.com/pod-product-compliance
Ingram Content Group UK Ltd.
Pitfield, Milton Keynes, MK11 3LW, UK
UKHW022314250926
475702UK00046B/779

9781032388878